AF529547

MUSHROOM
Cultivation in India

THE AUTHORS

Dr. B.C. Suman, Professor (Mushrooms), Dr. Y.S. Parmar, University of Horticulture and Forestry, Nauni (Solan), joined the University as Assistant Professor/Scientist in 1981. He did his Ph.D. in Mycology and Plant Pathology with specialization in mushroom research in 1989 and was actively engaged in teaching, research and extension activities in mushroom research and development. Dr. Suman has already published three books on mushroom and also more than 90 research/extension articles on various aspects of mushroom research in National and International Journal of repute. He was awarded by the University for doing meritorious work in the field of mushroom in 1997 with advance increments. His areas of specialization are selection and development of strains of *Agaricus bisporus* and standardization of parameters of spawn production technology.

Dr. V.P. Sharma, Senior Scientist, National Research Centre for Mushroom (ICAR), Chambaghat, Solan (HP) was born on 29th October, 1963 in Village Barin, Tehsil Sarkaghat, Distt Mandi (HP). He did his Ph.D. in Mycology and Plant Pathology (Mushrooms) in 1989. He joined Dr. Y.S. Parmar University of Horticulture and Forestry, Nauni as Assistant Mycologist in May, 1990. Dr Sharma served the university as Mycologist till February 2002. He was selected as Senior Scientist in ICAR and joined National Research Centre for Mushroom, Chambaghat in February 2002. Dr. Sharma has published more than 70 research papers on various aspects of mushroom research in National and International Journals. He was awarded Yadvindra Young Scientist Award twice in 1994 and consecutively in 1997 for doing best research in mushroom field. Dr. Sharma's main area of specialization are diseases and pests of mushrooms, spawn and cultivation of speciality mushrooms.

MUSHROOM Cultivation in India

B.C. Suman
Professor (Mushroom)
Spawn Production Laboratory
Chambaghat (Solan) – 172 213 (H.P.)
&
V.P. Sharma
Senior Scientist (Plant Pathology)
National Research Centre for Mushroom
Solan – 172 213 (H.P.)

2021

DAYA PUBLISHING HOUSE®
A Division of
Astral International Pvt. Ltd.
New Delhi - 110 002

© B.C. SUMAN (b. 1955–)
V.P. SHARMA (b. 1963–)

First Published, 2007
Reprinted, 2021
ISBN 978-81-7035-479-6 (Hardbound)

Publisher's note:

Every possible effort has been made to ensure that the information contained in this book is accurate at the time of going to press, and the publisher and author cannot accept responsibility for any errors or omissions, however caused. No responsibility for loss or damage occasioned to any person acting, or refraining from action, as a result of the material in this publication can be accepted by the editor, the publisher or the author. The Publisher is not associated with any product or vendor mentioned in the book. The contents of this work are intended to further general scientific research, understanding and discussion only. Readers should consult with a specialist where appropriate.

Every effort has been made to trace the owners of copyright material used in this book, if any. The author and the publisher will be grateful for any omission brought to their notice for acknowledgement in the future editions of the book.

All Rights reserved under International Copyright Conventions. No part of this publication may be reproduced, stored in a retrieval system, or transmitted in any form or by any means, electronic, mechanical, photocopying, recording or otherwise without the prior written consent of the publisher and the copyright owner.

Published by : **Daya Publishing House®**
A Division of
Astral International Pvt. Ltd.
– ISO 9001:2015 Certified Company –
4736/23, Ansari Road, Darya Ganj
New Delhi - 110 002
Phone: 011-4354 9197, 2327 8134
Fax: +91-11-2324 3060
E-mail: info@astralint.com
Website: www.astralint.com

Laser Typesetting : **Classic Computer Services**

Printed at : **Replika Press Pvt. Ltd.**

Preface

India is primarily an agriculture based country with about 1100 million human population and about 500 million live stock; 75 per cent human population lives in about 6,00,000 villages and is mainly engaged in agricultural and allied activities. Recently a lot of emphasis is given to the development of agro-based industry due to advantages of rural employment generation and gainful utilization of natural and farm resources. Large quantity of renewable lignocellulosic residues are generated every year as a result of extensive agriculture practices, the disposal of which creates a lot of problems. Mushroom cultivation is one of the largest economically viable commercial operation for bioconversion of lignocellulosic waste in highly acceptable nutritious food. Cultivating mushroom is an ideal income generating activity for landless labourers, unemployed youth and weaker section of the society of our country. It is also a perfect health food recommended for use to enrich diet with vegetable protein, vitamins, minerals and fibres. Moreover, the Indian sub-continent is endowed with a variety of vegetation and climate which are most suitable for growing different types of mushroom throughout the year.

Mushroom industry in India, is developing fast especially cultivation of white button mushroom. During the last 10-15 years, there has been perceptible change in the scenario, particularly in respect of production system, horizontal spread and vertical growth in productivity, with present production level touching 70,000 tons of all types of mushroom. The productivity of the button mushroom has also increased to 25-30 kg per quintal of compost.

That mushroom can serve as food is not itself of major importance, however, the way in which edible mushrooms can be grown greatly enhance their importance. The present book shows a great variety of cheap and useless materials, *e.g.* agricultural waste, industrial waste and forest waste that can be successfully used as media for growing the various edible mushrooms. The book emphasize mainly on the applied (cultivation) aspect of edible mushrooms and is intended for any one who is interested in edible mushrooms; the experienced mushroom specialist; seasoned commercial growers or simply the layman. The aim of writing this book, is that, it will encourage and stimulate further research on all aspects of edible mushroom with special attention directed towards discovering new edible species and improving both the quality and yield of existing ones. This would eventually lead to their mass production accompanied by reduction in cost, ultimately bringing this commodity with in the reach of common mass/ weaker section of the society.

Although every efforts has been made to improve the quality of book (in style and content) but there is always ample scope for improvement. Your critical appraisal as well as valuable suggestions regarding the improvement of book are always welcome.

B.C. Suman

V.P. Sharma

Contents

Chapter 1

Introduction

India is largely an agricultural country with cultivated area of about 2.5 per cent to feed almost 15.5 per cent of the world population. The emergence of food crises, due to an ever increase in the world population has led scientists and technologists to tap the available sources in finding alternative ways of producing cheap and quality food. Among the different food sources for humans, microbes is also a important source and in this fungi (edible fungi/mushrooms) comprises the largest and most important group. Mushrooms are the fruiting bodies of fungi, a group of organism, separate from plants and animals. According to Chang (1991) mushroom is defined as a macro fungus, with distinctive fruiting body which may either be above ground (epigeous) or below ground (hypogeous) and large enough to be seen by the naked eye and to be picked up by hand. In nature, mushroom grows wild in field, meadows and forests like, the common cultivated types. These constitute one of the major group of plant kingdom, with a difference that they do not possess green chlorophyll pigment and therefore, can not make use of the atmospheric carbon dioxide for synthesizing their own food. All the edible mushroom, under cultivation for the production of sporocarps without exception fall into either of the two major groups: Ascomycotina and Basidiomycotina.

The history of mushroom is as old as the origin of man himself. These have existed for millions of years and mankind has regarded their growth, both contemptuously and also as valuable food for flavour and nutrition. Theofastus (372-287 BC), the great Greek Philosopher wrote about food value of mushrooms during middle ages, when mushroom comprised royal dishes for the Greek and Roman Emperors. These have been described as 'Food for Gods' in classic writings of 'Vedas and Bible', while people in some other countries considered mushroom as devils creation.

In a botanical sense, mushrooms are fungi which lack chlorophyll and so cannot use solar energy to manufacture their own food as do green plants. However, mushroom can produce a wide range of enzymes that degrade the complex substrate, on which they grow, following which they absorb the soluble substances for their own nutrition. This absorptive nutrition is a characteristic of fungus (Chang, 1990 a). In the process, mushroom play an important role, in maintaining the cycle of nature by decomposing and clearing away left-over pieces of wood and other organic debris, not easily degradable.

In case of mushroom, the seed (spores) has no sexual origin (Atkin, 1981). The spores, which are invisible to naked eye, send out exploratory, root like fine threads (mycelium) in all directions, producing fruit body (the mushroom itself) without intermediate stem and leave stage. The mushroom fruit body may be umbrella like or of various other shapes, sizes and colours, but others may have additional structure like veil or annulus and a cup or volva. In contrast to green plants, most mushroom species are haploid. The diploid phase is normally transient and restricted to the basidium, where meiosis occurs. The meiotic products migrate in the basidiospore of which there are characteristically four per basidium. On account of such a life cycle and saprophytic nutritional habit, mushroom in nature are generally found to grow in places, where dead remains of plants or wooden logs are decaying. Although, majority of mushrooms are saprophytes, yet there are some, which either grow upon living plants as parasite or in symbiotic association with living plants, like plant roots as mycorrhizas.

According to a conservative estimate 1.5 million species of fungi actually exists and among these the number of known species of fungi is about 69,000 (Hawksworth, 1991). Out of 69,000 known and described species of fungi about 10,000 are fleshy macrofungi (Kendrick, 1985), of which half of them are edible (Chang 1993). About 2000 species from more than 30 genera are regarded as prime edible mushrooms, but only about 80 of them had been grown experimentally, 40 cultivated economically and about 20 cultivated commercially and 4 to 5 are produced on an industrial scale (Chang, 1990 b).

Domestication of the edible fungi dates back from centuries ago, but the most important development in the field have taken place mainly in the last 80 years. The cultivation accidentally originated sometimes in the middle of seventeenth century in France. French people cultivated mushrooms in caves and limestone pits, revolutionized the method for more scientific cultivation of *Agaricus bisporus* and remained as pioneer exporter of canned mushroom till recently. United Kingdom, United States of America, Japan, Italy and Netherlands picked up its cultivation more scientifically in the recent past. Korea, China, Germany, Belgium, Switzerland, Ireland, Rumania, Bulgaria, Poland and Russia are also endeavoring to find their place in the mushroom growing. In overall view, the world production of cultivated edible mushroom was 6,158 thousand tonnes in 1997 (Chang 1999) is now estimated to be over 12 million tonnes. It is mostly confined to Europe (55 per cent), North America (27 per cent) and Eastern Asia (14 per cent). Six major mushroom species *Agaricus bisporus, Lentinula edodes, Pleurotus* spp., *Volvariella volvacea, Flammulina velutipes* and *Auricularia* accounted for 87 per cent of the total world mushroom production (Chang, 1999).

The science of mushroom research and development in India, has expanded with passage of time. It is clearly evident from the fact that the yield of mushroom has increased manifolds during the last four decades *i.e.* from 2-3 kg/100 kg compost in 1964-65 to 18-20 kg per 100 kg compost now. *Agaricus bisporus*, popularly known as the white button mushroom, has the widest acceptability in India and its present production level is estimated between 60,000 to 70,000 tonnes per annum. The second most popular mushroom species, dhingri, accounts for approximately 7200–8500 tonnes per annum. However, this quantity is very meager, considering the huge market potential of this large and over populated country. In fact, in our country, there is need to convert the surplus agricultural waste (estimated to be over 18 crores tonnes) into nutritious food using simple and inexpensive methods. Cellulose, hemicellulose and lignin constitute about 80 per cent of total dry weight of straws. Most edible fungi possess good cellulose and hemicellulytic enzymes system and some in addition are highly lignolytic. Proper understanding and identification of such system will definitely help in selection of mushroom strains that have an increased ability to bioconvert more of the substrate into fruit bodies. Continued efforts in this direction seems to be bearing fruits, but the country has not been able to find its place, anywhere in the list of important mushroom growing countries of the world so far. Reasons have been several, but the most important one, has been the lack of proper education and dissemination of knowledge on mushroom growing. There are several gaps in our production technology also, resulting in low yields or crop failures. Moreover, major share of mushroom produce is still coming from cultivation under natural conditions by small farmers, where consistent yield are the major problems. Cultivation technology and strain developed in other countries for cultivation of mushroom under sophisticated controlled condition may not be suitable for our conditions. Scientists and technologists should address themselves in evolving hardy strains and production techniques most suited for our farmers and conditions. Besides, developing technology for big entrepreneurs, low cost, labour intensive technology should be developed for small farmers and landless labourers, which would go a long way, not only in generating employment for them, but also developing India into a leading mushroom producer in the world.

Chapter 2

Global Status of Mushroom Production

Cultivation of saprophytic edible fungi is an efficient way of utilization of cellulosic wastes. Mushrooms are delicious, nutritionally rich, medicinally important and non-conventional source of human food. Their cultivation technology is regarded as the second most important commercial microbial technology, next only to yeast for large scale profitable bioconversion of lignocellulosis agrowaste into valuable protein. Since mushroom consumption is popular all over the world, commercial cultivation is therefore equally important. The rapid rate of development of mushroom cultivation technology from primitive cave culture to a high technology industry during the last three centuries is a success story itself. Now, mushrooms are being cultivated in more than 100 countries of world, with an estimated total production of over 12 million tonnes.

Mushroom cultivation embodies the principles of microbiology, environmental technology and solid state fermentation in the conversion of domestic agricultural and industrial organic waste material into food for humans (Chang, 1991). The technique for cultivation of mushroom can be primitive as in rural farming of *Volvariella* and *Pleurotus* mushroom or can also be highly industrialized as in *Agaricus* and *Lentinula* production in urban areas. The present cultivation technology, therefore, includes a high degree of mechanization and ensures production round the year. The application of such technology has also helped significantly in maintaining a uniformity in mushroom production per unit area basis.

Presently there are 30 genera splitted up into 2000 species/ strains of edible fungi, of which cultivation of 6 to 7 has already assumed an industrial scale, while others have been exploited for expending cultivation in near future. Although many species are known to be edible, the production of cultivated mushroom is still dominated by *Agaricus bisporus* which accounts for approximately 32 per cent of the total world production (Chang, 1999). The other important species in terms of world output are *Lentinula edodes* (25.4 per cent) which is mainly cultivated in Japan followed by *Pleurotus* (14.2 per cent), *Auricularia* (7.9 per cent), *Flammulina* (4.6 per cent), *Volvariella* (3.01 per cent) and *Tremella* species (2.1 per cent).

Table 1: World Mushroom Production of Some Cultivated Edible Species

Species	*Fresh wt. (x 1,000 t)*				*Increase %*
	1986		*1997*		
A.bisporus/A. bitorquis	1227	(52.1%)	1956	(31.8%)	59.4
Lentinula edodes	314	14.4	1564	25.4	398.1
Pleurotus spp.	169	7.7	876	14.2	418.3
Auricularia spp.	119	5.5	485	7.9	307.6
Volvariella volvacea	118	8.2	181	3.01	1.7
Flammulina velutipes	100	4.6	285	4.6	130.0
Tremella spp.	110	1.8	130	2.1	225.0
Hypsizygus spp.	–	–	74	1.2	–
Pholiota spp.	25	1.1	56	0.9	124.0
Grifola frondosa	–	–	33	0.5	–
Others	10	0.5	518	8.4	5080.0
Total	2182	(100%)	6158	(100%)	182.2

Source: Chang, 1999.

The number of edible fungi, which have successfully been domesticated under sterile/unsterile condition is roughly forty and substrates used to cultivate the edible fungi can be placed into distinct categories.

1. Non exenic culture, with further divisions as:
   (*a*) Fungi which can utilize lignocellulosic wastes with little or pre-treatment, *Lentinula edodes* (Shiitake), *Auricularia* spp. (Jew's ear).
   (*b*) Those which can colonise plant material, *e.g.* straw after some composting, physical and chemical pre-treatment; *Volvariella volvacea* (paddy straw), *Pleurotus* spp. (oyster), *Stropharia rugoso-annulata* (giant mushroom), *Coprinus fimetarius* (ink cap).
   (*c*) Those which colonize plant material after a long composting process, *Agaricus bisporus/Agaricus bitorquis* (white button), *Coprinus comatus* and *Lepiota nuda*.
2. Axenic culture where sterilized substrate is used under controlled conditions:

   For economic reasons mushroom growers throughout the world need low cost preparatory technique that results in a substrate that can be used in a non-sterile environment.

Types of Mushroom Cultivated on Industrial Scale

Agaricus bisporus

Agaricus bisporus is predominately a crop of temperate region of the world and mostly confined to Europe, North America and Eastern Asia. Now it has also been successfully introduced in sub-tropical countries like Taiwan and Maxico. With the technical advancement during the last few decades, Taiwan became the largest single exporter in a decade. Korea, China and Indonesia which followed decade later are producing large quantities of this mushroom and earning foreign exchange for their countries while India is producing such small quantity that it does not find a mention in the world literature. Another species of this genus (*A. bitorquis*) was brought into culture during 1968 and since then attempts have been made to popularize and commercialize its production. *A.arvensis* and *A. subrufescens* seems quite promising which are being exploited for cultivation on industrial scale.

Lentinula edodes

This mushroom is commonly known as Shiitake in Japan and has enjoyed a good international reputation as food because of its distinct flavour. It has taken an important place in the agricultural industry of Japan, Korea and China, but other countries still depend for their supply from forests. Its commercial cultivation, in recent years has picked up tremendously and is the second most important species in terms of world output, contributing 25.4 per cent of the total world mushroom production (Chang, 1999).

Pleurotus Species

This mushroom is generally referred as oyster mushroom. It is a lignocellulolytic fungus and grows naturally in the temperate and tropical forest on dead and decaying wooden logs and sometimes on drying trunks of deciduous or coniferous woods. With more than hundred per cent biological efficiency coupled with its pleasant flavour and excellent drying and keeping qualities, it is likely to be a poor man's food. The oyster mushroom which belongs to the genus *Pleurotus* has many species of commercial significance *viz., P. ostreatus, P. sajor-caju, P. sapidus, P. florida, P. eryngii, P. cornucopiae, P. flabellatus, P. pulmonarius* and *P. eous*. This mushroom is only domesticated in Italy, Hungary, France, Germany, Taiwan, Japan, Korea, Thailand, India. Availability of large number of species of different colours, wide adaptability to temperature and ability to grow on variety of agricultural and forestry wastes, make this species most promising and its cultivation would go a long way in converting wastes into food and also helps in overcoming the problem of environmental pollution.

Volvariella volvacea

Volvariella volvacea is an edible mushroom of tropics and sub-tropics. It was first cultivated in China, during eighteenth century and subsequently introduced in South-East Asian countries. It is a fast growing mushroom with relatively simple cultivation technology involved. Its cultivation can be developed as cottage industry in both rural and sub-urban areas of tropical and sub-tropical regions of the world. *V. volvacea, V. diplasia, V. esculenta* and *V. bombycina* are some of the predominantly cultivated species on commercial scale.

Auricularia Species

Species of *Auricularia* are primarily grown in China and other Asian Countries, particularly Japan and Philippines. *Auricularia polytricha* is highly prized mushroom in Chinese market. Both species of *Auricularia viz. A.polytricha* and *A. auricula* are very popular dried mushroom in the world wide mushroom market, being delicacy and rich in nutrient. The optimum temperature for mycelial growth of cultivated *Auricularia* species ranges between 10-36°C but these grow better between 20-30°C and with 85 per cent relative humidity. A new method of cultivating *Auricularia* species in plastic bags has been developed and it is the mushroom which gives highest BE of more than 140 per cent.

Tremella fuciformis

Tremella species are known by different names such as jelly fungi, silver ear etc. It is generally cultivated on the wooden logs of hardwood trees like *Mangifera indica, Acacia confusa* and *Moris alba*. Taiwan and other Asian countries are known for its production. The dried fruit bodies are exported from Taiwan to South Asia, Japan and America. *Tremella* species are reported to have good medicinal value in curing several ailments such as tuberculosis and blood hypertension. The optimum temperature for mycelial growth is 25°C, whereas basidiocarp formation is favoured between 20-28°C with relative humidity around 90 percent. Light is considered essential for fruit body formation. It can also fructify on saw dust-rice bran (1 : 3 v/v) medium. A general view of the cultivation of some edible fungi in the world is given in Table 2.

Flammulina velutipes

This mushroom is commonly known as winter mushroom or Enoki and it ranks at number five in terms of world mushroom production. It is one of the most tastiest mushroom and grows in temperate climate. There is however a need to improve on many aspects of its culture. Yield and quality may be improved by breeding and selection. There is an urgent need to explore the possibility of utilizing other agricultural/forestry/industrial waste for its cultivation. The substrate presently in use are saw dust and rice bran only. It is sold fresh or dried.

Table 2: General View of the Cultivation of Some Edible Fungi in World

Species	Common Name	Temperature (°C)		Need for Light During Intensive Culture Lux)	Remarks
		Vegetative	Generative		
Agaricus bisporus	White button mushroom	22–25	15–17	0	Cultivated in more than 100 countries of world
A. bitorquis	Hot weather white button mushroom	28–30	24–26	0	Cultivated in Holland, UK and Belgium etc.
A. arvensis	–do–	24–26	28–30	50	Cultivated on experimental level
Pleurotus ostreatus	Oyster mushroom	24–25	10–15	200–500	Cultivated in Japan, France, Italy etc.
P. florida	*Pleurotus* 'Florida'	25	20–22	200–500	Cultivated in USA
P. cornucopiae	Branch oyster fungus	20	18–20	300–800	Cultivated in Asia and Europe
P. sajor–caju	Dhingri	25–27	25–26	200–500	–do–
P. eryngii	Dhingri (Kabul isolate)	25	18	200–500	Cultivated in Italy, France etc.
Volvariella volvacea	Paddy straw mushroom	32–37	35	300–1000	Cultivated in Asia
Pholiota nameko	'Nameko'	24–26	10–15	300–1000	–do–
Lentinula edodes	Shiitake	24	16–18	300–800	–do–
Stropharia rugoso annulata	Giant Mushroom	25	9–10	0	Cultivated in East Europe and Germany
Tremella fuciformis	Jelly fungus	25	22	100–300	Cultivated in Asia
Auricularia spp.	Jew's Ear	25–26	18–22	500–800	Cultivated in Germany
Coprinus comatus	Ink cap	25	18–28	0	Cultivated in Germany
Lepista nuda	Blewits	25	20–22	–	Cultivated in Netherlands
Flammulina velutipes	Winter mushroom	25	10–12	–	Cultivated in Asia
Tuber melanosporum	Truffle	25	22–25	–	Cultivated in France and Italy

Chapter 3

Importance of Mushroom

The land resources in the world for raising the food crops are limited. Net sown area generally fluctuates between 150-155 million ha. There is no possibility of its further increase. This warrants a wise use of land with due regards to its sustainability for long term productivity. Here, besides alternative crops which are good income yielders and more acceptable by farmers, the basic need is to tap other areas of economy which are found favourable with small farmers and landless labourer, who constitute one third of country's population. Apart from other areas, it is the commercial exploitation of mushroom, which promises a high return. India, with its great diversity of soil and climatic conditions can support to grow various edible mushrooms under natural condition in different parts of country.

Since, most of the crops are grown outside in the fields, they are subjected to the vagaries of nature. The mushrooms on the other hand, are grown indoors, where the environmental factors can be controlled to a fairly high degree by slight manipulation. We can raise five crops of mushroom in a year, at certain suitable locations. The yield of proteins from mushrooms per unit area is thousand times more than through raising the fodder crops for feeding animals for meat production. The protein deficiency is being felt all over the world and it is especially acute in the developing countries. To meet the needs of the growing populations, the scientist and technologist are looking for alternative sources of food among which mushroom yeasts and algal food are frequently mentioned. Except mushrooms, the other two are not liked because of their un-palatability.

Mushrooms being rich in protein, vitamins and minerals can form an important ingredient of food for a balanced diet. The nutrient absorbed by the mycelium, out of decaying organic matter accumulate and get transformed into various constituents of the mushroom fruit body. Mushrooms are thus rich source of nutrients, particularly proteins, unsaturated fatty acids, mineral, fibers and vitamins, such as B, C and D. Vitamin B, greatly helps in carbohydrate metabolism and is particularly useful in removing cardiac embarrassment and beriberi. Similarly, the presence of other vitamins like ascorbic acid, pantothenic acid, niacin and folic acid are useful in several other diseases. Their content of the anti-pellagra vitamin niacin is comparable to its level found in pork or beef, which are richest known source of this vitamin.

The mineral contents particularly calcium and phosporus are remarkably higher in mushroom than in many fresh fruits or vegetables, which again are extremely useful for body building processes. Most of the mushroom have very low starch content and form an ideal food for diabetic patients. More than their food value, the mushrooms are sought for their flavour and are used in soups, omletts or garnish for meat and other food. Their flavouring properties are well retained during canning, processing and cooking.

There is an immense potential for mushroom in India, where a very large population are vegetarian in habit. India, already has big name in markets of other countries of the world for its spices, condiments and mushroom. Mushrooms have very huge and wide scope for export as global market is expending very fast. Moreover, the per capita consumption of mushroom in countries like USA and Germany is very high per year. Thus mushrooms are a potential foreign exchange earner, which is much needed for the country.

In addition, mushroom industry is labour intensive and thus opens up new avenues of employment potential. A very large population in India is dependent on agriculture directly or indirectly. Mushroom cultivation can be taken up by the farmers as cottage industry and thus it can be a source of additional income.

The mushrooms have high medicinal values, the most important ones being their anti cancer, hypolipidemic, hypocholestermic and anti-hypertension effect. *Agaricus bisporus, Auricularia polytricha, Lentinula edodes* etc. have been known to lower down the blood cholestoral. *Calvatia gigantea* and *Lentinula edodes* produces high molecular weight substances which inhibit both influenza and polio-viruses. Calvacin, a potential anti-tumor agent has been isolated from the giant puffboll, *Calvatia gigantea* which can be used against cancer. Antitumour activity has also been found in the Maitake Mushroom (*Grifola frondosa*). Water extract of this mushroom has recently been reported to kill the HIV virus causing AIDS in human. *Ganoderma lucidum,* the Reishi Mushroom is the most important medicine in China and has been used for wide range of health benefits from preventive measure to maintenance of health (being rich in high molecular wt. polysaccharides, triterpens, Ling Zhi-8, adenosine derivatives and organic germanium GE-132). It is perhaps the only mushroom today which is being used for preparing pills for marketing in several countries including India.

In addition to their fascination for gourmet and medicinal properties, mushrooms are also utilized for making various articles. The fruit bodies of *Fomes fomentarius* and *Ganoderma applanatum* are used to produce suede-like material from which hats, various articles of dress, handbags and picture frames are made. Dried *Cariolus varsicolor* brackets has been used for making hats for costume decoration while bottle corks are made from *Polyporus squamosus, P. applantus* is used for curio and also for purpose of etching. *Daedalea quercina* is sometimes used to clean down horses particularly those with tender skin. It is also used for men, for cleaning hair. Tinder mushroom *Polyporus fimentarius* or tinder mushroom is used in manufacturing of tinder. *Polyporus nigricans* when dried and pounded is an ingredient in snuff. *Polyporus bispidus,* which gives a brown dye, is used for colouring silk, cotton and wool. This is also used for leather dresser to give an chestnut colour and carpenter to give brown colour furniture. *Polyporus sulphureus* gives yellow colour and *Fomes ignitarius* gives a brown black colour. *Coprinus comatus* is very deliquescent and soon becomes black liquid, which can be used for writing purposes.

Similarly shaped fruit bodies of *Polyporus fomentarius* and *P. ignitarius* are used for flower pots. The tubed portion is cut out and hoofed portion is inverted and hanged. Many edible fungi are luminescent. The decayed wood when permeated with mycelium of *Armilearia mellea* glows strongly, as long as growth continues and it remains moist. The fruits bodies and mycelium of *Fomes anosus, Pleurotus janopicus, Boletus edulis* and *Callobia longipes* etc. emits light and luminosity is often so bright that one can read words of printed pages in dark.

Thus, there are various extremely important roles that mushrooms play in the world. Their usefulness to man as food, as tonics and medicinal and also in the bioconversion of waste organic materials to forms that can enter the major nutrient cycles are all of great benefit to both man and nature.

Chapter 4

History of Mushroom Cultivation in India

Though mushroom cultivation both in east and west started many centuries ago; yet its cultivation in India, is of recent origin. It was probably in the year 1886, when some specimens of mushroom were grown by N.W.Newton and exhibited at the Annual Show of Agriculture, Horticulture Society of India. In the year, 1896-97, Dr.B.C.Roy of the Calcutta Medical College, carried out chemical analysis of the local mushroom prevalent in caves or mines. Thorough search of edible mushroom was instituted by Sir David Prain in the year 1908. Professor S.R.Bose of Bengal, was successful in culturing two agarics on a sterilized dung media in the year 1921.

Paddy straw mushroom cultivation was first attempted in India at Coimbatore, in 1943 by Thomas and his associates. However, first systematic attempt in cultivating button mushroom was made in 1961, when a scheme entitled "Development of Mushroom Cultivation in Himachal Pradesh" was started at Solan (H.P.) by Himachal Pradesh Govt. in collaboration with ICAR, New Delhi. In 1964, cultivation of *Agaricus bisporus* on experimental basis was started by CSIR and State Govt. at Srinagar in Jammu and Kashmir. In the late sixties, a few progressive growers in Himachal Pradesh and Jammu and Kashmir started growing button mushroom on commercial scale *viz.* Teg's Mushroom, Chail (HP); Saigal Mushroom Farm, Kasauli(HP); Harco's Srinagar(JK) and Col.Kak's Mushroom Farm Srinagar(JK) (Dhar, 1997). In 1965, Dr.E.F.K.Mantel, F.A.O. Mushroom Expert, guided and assisted, Department of Agriculture for construction of a modern spawn laboratory and fully air-conditioned mushroom house at Solan (H.P.). Research on evaluation of different strains and use of various agricultural wastes and organic manures and fertilizers for preparing synthetic compost were undertaken. In the early seventies, mushroom cultivation started spreading to other hilly regions of Uttar Pradesh and Tamil Nadu.

In the year 1974, Dr.W.A.Hayes, another F.A.O. Mushroom Expert, guided further in improving the method of compost preparation, pasteurization and management of important parameters in the mushroom house at Mushroom Research Laboratory, College of Agriculture, Solan(HP). New Compost formulation, casing materials and important parameters like nitrogen content in the compost, moisture in casing soil, air movement and maintenance of proper environmental factors were also standardized which raised the mushroom yield from 7 to 14 kg per 100 kg compost. Solan in Himachal Pradesh was developed as a nucleus centre for intensive research on all aspect of cultivated mushroom including development of spawn and spawning technique and compost and composting technology. Research for identifying proper casing soil mixture from available raw material under local conditions, management of important insect-pests, diseases and nematodes, as well as to meet other requirements were also initiated at Mushroom Research Laboratory of the Department of Mycology and Plant Pathology, College of Agriculture, Solan earlier affiliated to H.P.K.V.V. Palampur and later to Dr.Y.S.Parmar University of Horticulture and Forestry, Nauni (HP), under the dynamic leadership of Dr.R.L.Munjal.

The initial success in cultivating *Agaricus bisporus* by employing the technology developed indigenously, the project influence spread all over the country. Indian Council of Agricultural Research, therefore, considered the importance for establishing a National

Mushroom Research and Training Centre at Solan, sometimes during Oct., 1982 with seven sub- centres at Ludhiana, Kanpur, Pantnagar, Kalyani, Pune, Bangalore and Srinagar. National Centre for Mushroom Research and Training started functioning w.e.f. 8.6.1983. Simultaneously, Indian Council of Agricultural Research sanctioned six All India Coordinated Mushroom Improvement Projects in the year 1983 during VIth Five Year Plan at G.B.Pant Univ. of Agril. and Technology, Pantnagar(UP), Punjab Agricultural University, Ludhiana(Punjab), Tamil Nadu Agricultural University, Coimbatore(TN), Bidhan Chandra Krishi Vishwa Vidyalaya, Kalyani(WB), Mahatma Phule Agricultural University, College of Agril. Pune(MS) and C.S.Azad University of Agril. and Technology, Kanpur(UP). In subsequent years Kanpur and Kalyani centres were deleted and I.G.K.V.V. Raipur(Chattisgarh) was added.

With the concerted efforts of all those engaged in mushroom research and development, particularly the mushroom scientists, the yield of mushroom in India, in the recent years has increased manifolds *i.e.* from 2-3 kg per 100 kg compost to 18-20 kg per 100 kg compost and its present production level is estimated between 60,000 to 70,000 tonnes per annum. The current Indian scenario is quite impressive, but the country has not been able to find its place anywhere in the list of important mushroom growing countries of the world so far. Reasons have been many, but most important one has been the lack of proper education and dissemination of knowledge on mushroom growing. There are several gaps in our production technology resulting in low yields and crop failures. The mushroom scientist therefore, should develop the cultivation technology most suited to our farmers and according to the prevailing environmental conditions here.

Chapter 5

Food Value of Mushrooms

From ancient times, the mushroom has been consumed as a food by man. Initially, it was probably the pleasing flavour and texture that was so attractive and in some societies their use was limited to royalty but for well over two thousand eight hundred years, some species of mushrooms have also been used as medicinal and tonics. Mushrooms are the most exquisite of the gastronomic treasures of the world. They have been prized as the food of the gods on account of their special flavour, nutritional value and medicinal properties. Mushrooms blend well with most of the vegetables and spices to form delicious items of food.

Recent technological breakthrough in crop productivity coupled with increase in area under cultivation have led to a tremendous increase in production of not only grains, but also of crop residues. It has been estimated that out of 38 crores tonnes, around half of the total quantity of plant residue from agricultural and industrial process remain unused. Main crop residues in India are cereal straws, sugarcane bagasse and leaves cotton waste and wood dust, coconut coir waste and rice husk. Burning of crop residue *in situ*, besides posing environmental problem is a terrible waste of energy and very poor utilization of crop residues. Allowing it to decay in the field, helps in perpetuation of various pests and pathogens along with the build-up of various saprophytic moulds responsible for health hazards. These crop residues could be converted into food, feed and fuel by mushroom cultivation. This bioconversion of insoluble plant waste materials to a nutritious food that has a high protein is one of the important activities of mushrooms.

Although, systematic cultivation of mushroom started during the reign of Louis XIV, yet even before that people used to collect mushroom from wild and consume them. Mushrooms have traditionally been used in India as garnishing on food and few varieties have also been treated as tonic or medicinal material. Sincere efforts to bring mushroom under cultivation were made in 1961 after the commissioning of a scheme "Development of Mushroom Cultivation in India", at Solan in Himachal Pradesh. But now, with the availability of choice species vulnerable for cultivation mushroom, which are considered as delicacy, has not become the food of common man. Mushrooms are rich in proteins, vitamins and minerals, while poor in fat and carbohydrate. The mushroom proteins are considered to be of high quality containing all the essential amino acids required for an adult individual. Some of the amino acids like tryptophan and lysine, which are absent in vegetable protein are present in mushroom protein. They are also known to be excellent source of riboflavin, nicotinic acid and pantothanic acid. Mushroom contains appreciable amounts of vitamins, B, C and D. Their content of the anti-pellagra vitamin niacin is comparable to its levels found in pork or beef, which are the richest source of this vitamin. Minerals, such as, calcium, phosphorus and potassium are present along with a fair quantity of copper and iron, all of which are essential for the formation of bones, teeth and also for good eye sight. Mushrooms are very low in sodium, while they are good source of potassium, a mineral which help balance the sodium, which comes from salt. A brief account of its nutritive value is given in the following text.

Protein is the most critical component contributing to the nutritional value of a food. Edible mushrooms are considered as a potential substitute of muscle protein on account of their;

(i) digestibility (89 per cent), (ii) good amino acid content and (iii) about 1000 times higher production of mushroom protein per unit area. Mushroom contains 20-35 per cent protein (on dry weight basis) which is higher than in vegetables and fruits. Mushroom protein is also of very good quality and can be a substitute for animal protein. It also contains most of the amino acids in sufficient quantity and can prove a good supplement to those cereal diets, which lack in two essential amino acids, namely lysine and tryptophan. Methionine and cystine, the two essential amino acids lacking in mushroom is compensated by cereals in India diet. There is not any deficiency of calories in Indian diet, except quality protein. Moreover, protein malnutrition particularly in infants, children, pregnant and lactating women is the biggest nutritional problem of our country. Keeping in view, the declining per capita availability of pulses, mushroom fit in very well in the diet of predominantly vegetarian population of our country for maintaining a proper protein balanced diet. Mushrooms have been recognized by the F.A.O. as food constituting to the protein nutrition of the countries depending largely on cereals.

Mushrooms appear to be good source of several vitamins including thiamine (B_1), riboflavin (B_2), niacin, biotin and ascorbic acid (Giacomini, 1957), vitamin A (retinal) activity is relatively uncommon although several mushroom contain detectable amounts of pro-vitamin-A, measured as L-carotene. Similarly vitamin-D, activity is rare in mushroom, but may contain ergosterol, which can be converted to vitamin-D, under ultraviolet irradiation (Rambottom, 1953). Folic acid and vitamin $B_{12,}$ which are so much needed by pregnant and lactating women and are almost absent in vegetables food, are present in mushrooms. They are also a fairly good source of vitamin C (4-8 mg/100gm). In *Agaricus bisporus*, vitamin K and vitamin E have also been reported.

Mushrooms probably contain every mineral present in their growth substrate and are good source of them. In general, mushroom contains significant quantities of phosphorus, sodium and potassium, a lesser amount of calcium and a very low amount of iron. Though iron is in low quantity, but it is present in available form and has been shown to maintain blood hemoglobin level. As compared to other food substances potassium/sodium ratio in mushroom is very high which is desirable for the patient of hypertension. Fresh specimens of mushroom may contain 0.03–13.5 ppm lead, 0.02 per cent ppm cadmium and traces of selenium (Leh, 1975).

The fat content of mushroom particularly *A. bisporus* is very low (0.3 per cent), but rich in palmitic, stearic, oleic and linoleic acid (Holtz and Schisler, 1971). Linoleic acid is especially abundant comprising approximately 70 per cent of the fatty acid content of natural lipid fraction. This high linoleic acid concentration is comparable to that found in safflower seed oil (Summer, 1973). Cholesterol the dreaded sterol for heart patient is absent and in its place ergosterol is present in concentration of 0.2–270 mg/100g dry weight (Geocomini, 1955), which can be converted into vitamin D by the human body.

Total carbohydrate content is 4–5 per cent which consists of chitin, hemicellulose and glycogen. *Agaricus bisporus* consists pentoses, methyl pentoses, hexoses, disachharides, amino sugars, sugar alcohols as well as unidentified uronides and methyl sugars (Holtz, 1971). Mannitol appears to be in particularly high in concentration (9–13 per cent). Very little amount of free sugars (0.5 per cent) are present. Starch is absent. Mushroom are low caloric food with very little fat and highly suitable for obese persons. Fibre content is very high (3–32 per cent) which is helpful in excretion of waste and prevention of constipation.

On an average, fresh mushrooms contain approximately 90 per cent moisture and air dried mushroom 10–12 per cent. This high moisture content is not unusual, however, if mushrooms are compared with other common vegetables. Fresh specimens of truffles (*Tuber* spp.) generally contain about 80 per cent moisture (Barbero, 1969) and mushroom species having a leathery texture might be expected, also to have a lower moisture content.

On the basis of all those qualities the nutritional value of mushrooms as compared to other food-stuffs appears to be quite high. Proximate composition of the cultivated white button mushroom (*Agaricus bisporus*) is given in Table 3.

Mushroom yeast and algal food are frequently mentioned as alternative source of protein/food. Of these mushroom are the most preferred since they offer greater potential (160–180 kg/m^2 production annually from 5-6 crops of white buttons). Mushroom contain 20–35 per cent protein (about 3 per cent on fresh weight basis) which is higher than most of the vegetables. Quality of mushroom protein is superior to vegetable protein and as good as animal protein.

Table 3: Proximate Composition of *Agaricus bisporus*

Proximate Composition (% Fresh Weight)		*Essential Amino Acids (% Crude Protein)*	*Fatty Acid Composition (% Total Fatty Acid)*	
Moisture	90.1	Leucine	Saturated	19.5
Protein	3.0	Isoleucine	Unsaturated	80.5
Fat	0.3	Valine	Linoleic acid	69.2
Carbohydrate	5.0	Tryptophan		
Fiber	0.9	Lysine		
Ash	0.8	Threonine		
Calorie	36.0	Phenylalanine		
		Methionine		
		Histidine		

Digestibility of mushroom protein is also high. Due to low calories, low fat (rich in linoleic acid and lacking in cholesterol and high K/Na ratio, mushrooms are dieticians choice for those with obesity, hypertension and heart diseases. With low calories, no sugar and starch and high protein content, they are said to be the delight of the diabetic. Alkaline, ash content coupled with high fiber content present in mushroom is suited to those with hyperacidity and constipation. To correct the irregularities of carbohydrate metabolism, cardiac discomfort and Beri-Beri disease, mushroom are highly recommended due to their high content of thiamine. Scurby disease can also be cured to some extent by eating mushroom a misconception has crept into the minds of people, that they are more nutritive than other mushrooms. *Pleurotus* and *Volvariella* are also as nutritious as button mushroom. Rather, their cultivation should be encouraged in our country because of the existing agro-climatic conditions and their utilization of wide varieties of agricultural and forestry waste and easy method of cultivation. Small farmer and landless workers should be motivated to cultivate/ produce and consume more mushrooms.

Chapter 6

Steps in Mushroom Growing

Mushroom growing consists of the following steps;

1. Selection of site and types of mushroom to be grown
2. Mushroom farm structure, design and layout
3. Principles and techniques of compost and composting process
4. Principles and techniques of spawn production
5. Spawning and methods of spawning
6. Casing and crop production
7. Environmental crop management
8. Harvesting and marketing

(i) Selection of Site and Types of Mushroom to be Grown

While selecting a site for establishment of a mushroom unit in a particular region, some basic factors have to be taken into consideration which will decide the size and capacity of farm. First and foremost is the existing environmental and agro-climatic condition of the area, specially with relation to the types of mushroom to be grown. For example, for raising crop of white button mushroom (*A. bisporus*) it is not advisable to build a farm in region with temperature remaining above 25°C for long period of time, otherwise the cost of insulation and energy will take a heavy toll. However, exception must be made in case of *Agaricus bitorquis*/*Pleurotus* species which grows best at temperature between 25-30°C and *Volvariella volvacea*/*Calocybe indica* which grows best above 30°C temperature. There is an immense potential for mushroom cultivation in India, because of varied agro-climatic condition prevailing at different geographical regions of the country. Moreover, large quantities of forestry, agricultural and industrial wastes are produced in our country every year, causing a great problem of their disposal. In case a small quantity of these waste products could be utilized for cultivation of mushroom particularly the species of *Pleurotus*/*Volvariella*, not only the problem of their disposal will be solved, but it will be converted into a very nutritious food.

Second most important factors is that the grower must find out, whether enough raw material and cheap man power are available in the area and whether there is a ready market for their produce. It would be of great advantage to build a mushroom unit in a region, where there are already some units in operation or is in developing stage. This will help the grower in procuring all kind of raw material (required for compost and spawn preparation) as well as guidance from technical expert/advisors as and when required.

Moreover, mushroom being an indoor crop, does not require arable land, except for some non-agricultural land to build the infrastructure. The farm should be build on the site in such a way that the compost yard and dumping quarters for supply and removal of compost, casing soil etc. are easily accessible for heavy truck. In this regard, care must be taken that the fresh manure and casing soil should not come into direct contact with spent compost and that, if possible the growing rooms should be separated from that part of farm, where pasteurization and spawn run take place. In addition, sufficient space should be kept for future expansion.

(ii) Mushroom Farm Structure, Design and Layout

Mushroom being an indoor crop, does not require arable land except for some non-agricultural land to build infrastructure for preparation of substrate, raising of crop, preparation of spawn and post harvest handling etc. White button mushroom in India, is grown seasonally, as well as in environmentally controlled cropping houses and both require building of basic infrastructure. Therefore, in mushroom growing, one of the most important conditions for obtaining good results at low cost price is an adequately built and properly equipped farm. An ideal and profitable mushroom, consists of a number of buildings, constructed on a piece of land for specific purposes, needed for growing mushroom for profit. For raising of these infrastructure will require at least one hectare of land, taking also into consideration, future expansion. The land should be near to the highways and well connected by approach road of at least 4-5 meter of width with sufficient space for turning of vehicle. The land should have permanent and regular supply of clean water, availability of 3-phase electricity connection and proper drainage system.

An ideal and profitable mushroom unit should have at least 10 growing rooms of 900 sq. feet bed area each with capacity to have 10-12 tonnes of ready compost at one time. For pre-wetting and phase-I of composting, the composting yard should be of dimension of 60′(l) × 40′(b) × 20′(h) with G.I. roofing and covered from all sides upto 2-3 feet high brick wall to avoid compost preparation process from rain water, strong sun light and other environmental hazards. There should be regular supply of clean water at composting yard with proper drainage system. In principle, two systems of growing are recognized in the field of mushroom growing *viz.*, one zone and two zone system with modification to suit local condition. When one zone system is used, pasteurization, spawn running and growing is done in one room, whereas in two zone system, there are separate rooms for pasteurization, spawn running and cropping.

In India, mushroom cultivation is of recent origin and protected cropping of mushroom was started on a small scale in 1961 at Solan (H.P.) and later it spread to other hilly regions of neighboring state of Jammu and Kashmir, Uttar Pradesh and Haryana as cottage industry. Earlier, button mushroom cultivation was done in already built structure with minor modification for ventilation and exhaust vents etc. The compost was prepared by long method of composting and growing was done in wooden trays or on shelves. However, with the establishment of Mushroom Research Laboratory of Department of Mycology and Plant Pathology, College of Agriculture, Solan (H.P.) and Mother Composting Unit of Department of Horticulture (H.P.) a large number of mushroom farms were established in and round Solan areas. With the advancement in growing technology and technical know-how provided by the Scientist, now environmentally controlled mushroom houses are being established in different agro-climatic zone of India. More and more, latest modification and innovation are perfected to suit growing condition in India and mushroom farm design developed using locally available materials/machinery for higher productivity at lower costs. Following the Dutch pattern of bulk pasteurization in tunnels and growing in beds, laid in shelves, inside the environment controlled cropping rooms, the mushroom industry received a great boost in the last 15-20 years in India. The importance of proper and suitable design for commercial mushroom farm, suited to conditions in India, has become a necessity now and needs to be done with skill and under expert's guidance. A layout plan of a mushroom unit suiting Indian conditions is given in Figure 1.

Selection of Site

The following points should be taken into consideration before setting a mushroom farm, in a particular area:

1. The site should be nearer to highways or well connected by approach road of at least 5 meter of width with sufficient space for turning of vehicle.
2. Availability of permanent water supply.
3. Availability of permanent electricity supply with three phase electric connection.
4. Availability of cheap labour and raw materials.
5. Near to market or city with sufficient space for future expansion.
6. Proper drainage system.

After proper selection, the construction and development of infrastructure should be done in a planned and phased manner. An ideal mushroom unit should have the following structure.

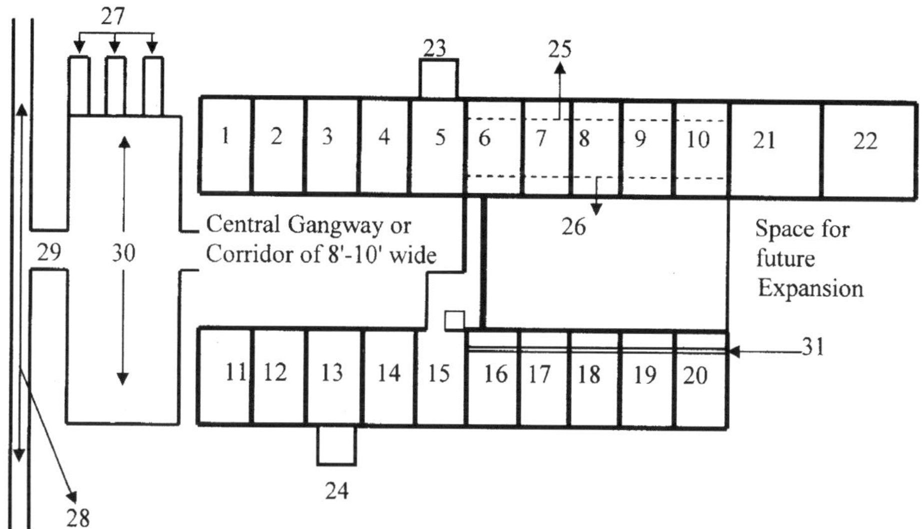

All rooms are insulated with thermocol insulation material and are of 16.6' x 10.10' in size

Room No. 1 and 2: Pasteurization room/tunnel; Room No. 3 and 13: Spawn running room; Room No. 4: Store for implements; Room No. 5: Cold Store for fresh mushroom; Room No. 6 to 10 and Room No. 16 to 20: Cropping rooms (45' × 20' × 10'); Room No. 11: Field laboratory; Room No. 12: Glass ware room; Room No. 14: Boiler room; Room No. 15: Machine room (plant room) plant for air conditioner for Room No. 6 to 10 and 16 to 20; 21: Post harvest handling unit; 22: Spawn Laboratory; 23: Machine room for Room No. 5; 24: Machine room for Room No. 3 and 13; 25: Under ground masonry return air duct; 26: Insulated masonry supply air duct on roof; 27: Overhead water tanks; 28: Pathways; 29: Access road; 30: Composting yard; 31: Air Duct.

Figure 1: General Layout of a Mushroom Unit

A. Composting yard	For preparation of compost by long method, for phase-I of composting (short method), pre-wetting and outdoor composting.
B. Bulk Chamber/ Pasteurization tunnel	For phase-II of composting (short method, bulk method).
C. Casing pasteurization chamber	For pasteurization of casing materials.
D. Cropping Rooms	For raising crops (Natural condition/Environmentally controlled)
E. Spawn Laboratory	For spawn preparation.
F. Post harvest handling unit	For preservation, canning and packaging of mushrooms.
G. A.C. Room	For housing compressors, cooling tanks, motors etc.
H. Boiler Room	For housing steam boiler.
I. Generator Room	For housing generator
J. Office/Store Room	For office purposes and also storing various ingredients of compost and spawn.

In addition, some other infrastructure can be raised depending upon the requirement of a particular farm. The composting yard should be built near the approach road, whereas pasteurization/ bulk chamber should be built on the other side of the composting yard, so that the distant end of the chamber opens nearer to the cropping room and away from composting yard. The casing pasteurization chamber should also be build away from composting yard or on one side of the bulk chambers. An ideal and profitable mushroom unit should have at least ten growing rooms of 900 sq. feet or 100 sq. meter bed areas each with possibilities for future expansion, build on both sides of a common corridor of at least 15-20 feet wide. A wide variety of design and construction materials have been used for building mushroom houses. The early houses usually had a pitched roof and a flat false ceiling, which was insulated at first with granular cord, later with fiber glass, rock wool, verculite or foamed plastic. A very popular form of construction was the house made of curved corrugated asbestos sheet. The Dutch farm design and construction are widely being adopted all over the world with suitable modification as per local conditions (Veddar, 1978), Shandilya (1982) conductive comparative study for growing mushroom in an insulated mushroom house and in mud houses, especially constructed for this purpose.

Construction of Infrastructure

Covered Composting Yard

For preparation of compost by long method or for phase-I of composting (short method) *viz.*, pre-wetting and outdoor composting, composting yard is required preferably without side walls or partially surrounded by 2-3 feet high brick wall to avoid rain water interference in the normal process of composting. The compost yard should be approachable by a motorable road, so that the raw material can directly be unloaded on one end of the platform. A composting yard of 60′ × 40′ × 20′(h) will be required for preparation of 12-15 tonnes of ready compost, sufficient for a growing rooms of 45′ × 20′ × 10′(h). Such yards should be cemented with drains for carrying away the liquid leaking from the manure. The drains often should lead to a common tank (guddy pit) so that liquid collected can be sprayed over the material again. The guddy pit should be provided with dewatering pump and hoze. The yard should be covered with GI roofing to protect the compost piles against strong sun light, dehydration, rain, snow and strong winds. The composting yard should be constructed on tresses or RCC pillars 20 ft high. The foundation of composting yard should be properly made on firm ground. A foundation of 2 feet is dug first and then filled with sand upto 8" followed by another layer of broken stone upto another 8" and finally with 8" layer of cemented floor, preferably as reinforced cement concrete. Provision for underground aeration is made by laying perforated pipes connected to a blower. The floor is given a run-off of 1 cm per running meter, away from the bulk chamber and towards the guddy pit. A permanent 3-phase electricity connection and regular water supply should be available on the yard with properly laid drainage system. The yard should have proper light system, to facilitate round the clock operation at the composting yard. An overhead water tank is necessary, particularly where water is scarce, to store water for carrying uninterrupted and timely operation.

Pasteurization Chamber/Tunnel

Since pasteurization chamber/tunnel is used for phase-II of composting *i.e.* for pasteurization and conditioning, therefore, a reinforced concrete foundation is recommended with insulation. Two types of chambers are built for this purposes, that is;

1. Peak-heating chamber and
2. Bulk-pasteurization tunnel

Peak Heating Chamber

Also called as single zone system and consists of an insulated room with facilities for injection of steam, air intake and recirculation. All the operations *viz.*, pasteurization, conditioning, spawn running, casing and cropping is done in the same chamber for labour economy and efficient utilization of the available space. In this system all rooms, on a mushroom farm are preferably insulated and provided with steam, air handling, cooling and heating facilities. The compost after phase-I is filled in racks/trays in a cropping rooms for all operations in a row (Veddar, 1978). This system can be more efficient in already built structures like cold storage etc. where the entire space is utilized to maximum efficiency without investing in construction of bulk chambers for composting.

Bulk Pasteurization Chamber/Tunnel

Also called as double zone system in which the compost after phase-I is filled into specially built chamber, which is excellently insulated and provided with steam connection and air blowing system for recirculation (Figure 2).

The dimensions for a tunnel to be used for the mass treatment of compost depend on the quantity to be filled. Most commonly used tunnels are of the dimensions of 45′ × 8′ × 13′ (h). The thickness of beds of compost can vary according to the density, which depends on raw content, straw hardness, quantity of animal excrement or other animal matter, moisture content, degree of previous composting. Also because during peak heating about 25 per cent of the compost is lost and about 7 per cent during spawn running (loss of moisture and dry matter) the tunnel must therefore be filled with 30 per cent more of compost before pasteurization. In a layer of approximately 6 to 6-7 feet, 1 ton to compost needs a floor area of about 3.7 to 4 sq. feet. When the layer are deeper, the composting especially the lower part may be sagging too closely (Tunney, 1978).

The floor of the bulk chamber is a double one. The lower floor of isolated concrete lies about half a meter below level. The second

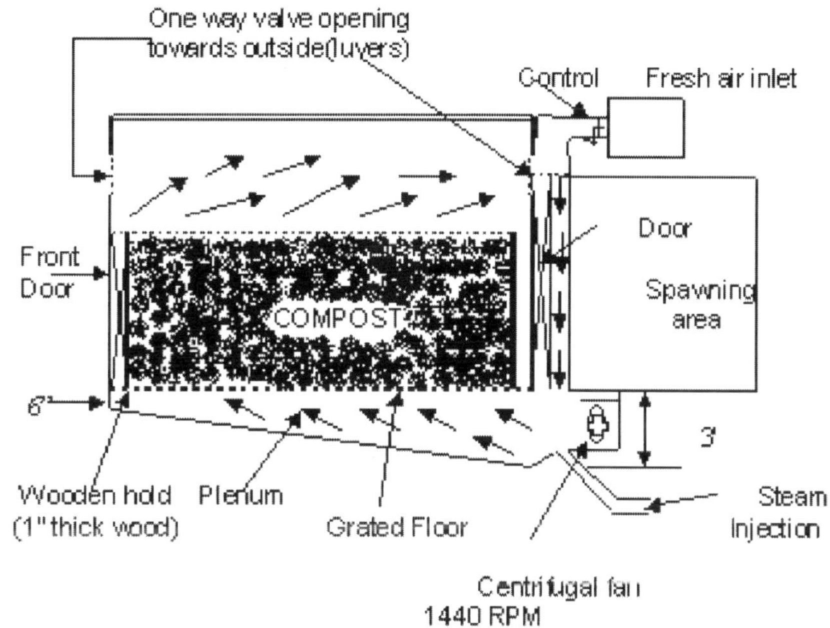

Figure 2: Cross Section of Compost Bulk Pasteurization Chamber (L 45' × B 8' × H 10')

floor consists of wooden beam with opening between (20-25 per cent) or cattle grids. Under the floor should be sufficient air circulation, the free space between the two floors should, therefore be at least about 50 to 60 cm. for a good pressure distribution and no restriction of air velocity, it would be desirable if atleast 2 per cent of the floor could slope down to the air inlet. At that lower point, provisions are made to carry off the leaking and the condensed water (Tunney, 1978). The bulk chamber should be constructed on one end (away from road) of the composting platform. One end of the bulk chamber should open into the platform and the distant end in the clean spawning area. The foundation of bulk chamber or peak heating rooms should be strong enough preferably on firm base/ ground. A well insulated floor can be obtained by spreading a layer of plastic sheet on a smooth layer of sand of about 8". On top of this is poured a work floor of 2" thickness of concrete. After hardening the floor is covered with sheets of materials containing asbestos or thermocol sheets. For heating system in the floor, the plastic pipe are laid first, then concrete flooring carefully poured over them to a thickness of about 3-4". In this way, a proper insulation is done which does not differ from the walls of the chamber. The walls should be 9" thick (one brick length) built over the concrete foundation. The roof is made of 4" thick RCC. The walls, ceiling and the floor below the plenum are insulated with 5 cm thick insulating material (15 kg density per cubic meter) necessary for effective insulating effect during pasteurization and conditioning of the compost (Veddar, 1978). The grated floor preferably of square wood of 1-1.5" size should be laid, above the plenum over the ventilation duct. Space of 20–25 per cent should be left in grated floor for proper air passage (ventilation/ circulation).

The walls of bulk chambers should have a very good co-efficient of insulation. A, K-value of about 0.3 is generally highly suitable. The co-efficient K is the quantity of heat passing through a wall, expressed in K cal/sq. meter, per hour and per degree of difference in temperature. The floor of the room must be carefully water proofed with three coats of bitumen or by means of suitable plastic sheeting applied to the surfaces. To keep the insulation, materials dry, the inside wall should be given a vapour proof finish with a coating of bituminous emulsion. The roof should be made with 4" thick RCC and covered with corrugated asbestos cement roof sheets which are attached to roof anchor (Veddar, 1978). One or both of the short sides of the bulk chamber should have double doors which must close hermitically and properly insulated with 2–3" insulation material in the centre and covered with aluminum sheet on both sides. The doors on the back wall of room should be 8" high to facilitate the filling with slated conveyor belt. A second wooden wall of 6" thickness can be constructed behind the doors to prevent the compost from pressing against them. There should be two exhaust vents in the chamber, one for re-circulation exists and other for exhausts of gases on introduction of fresh air by dampers. The pasteurization chamber should be fitted with a blower fan of 5–7.5 H.P. motor, below the plenum, with a speed of 1440 rpm corresponding to air pressure supporting 120-121 mm water pressure *i.e.* 5 inches W.P.G. Based on the number of doors, the tunnel can be of two types *i.e.* single door and double door tunnels. In single door tunnel obviously, the same door will be used for filling and emptying, whereas in two doors bulk chambers, one door will be used for filling and other for emptying.

Casing Pasteurization Chamber

The casing pasteurization chamber is also an insulated chamber with a steam connection and a blower for effective circulation of steam inside the chamber for attaining correct temperatures for pasteurization of the casing materials. The walls of casing chamber should be made of cavity walls of concrete 8″ thick, with an air space of 2″ between them. The dimensions of chamber will depend on the quantity of the casing soil that must be steamed. On the inner walls, asbestos cement sheet will be placed crosswise on supports. After this, a layer of 16″ thick foam glass or 6″ thick rock wool is spread on ceiling. The grated floor inside the chamber on which casing soil will be placed in trays should be 10″ above the concrete floor and is made of strong wooden boards arranged like grate on supporting beam. Under this grated floor steam connection will be provided with blower for effective circulation of steam. Proper arrangement for drainage should be provided in the concrete floor. The door of the casing pasteurization room should also be properly insulated as in compost bulk pasteurization chamber and made air tight by fixing rubber gasket on the inner side of the door.

Cropping Rooms

Two types of cropping rooms can be constructed depending upon the requirement of the area, where farm is located viz;

1. Natural Environmental Condition Growing Rooms.
2. Environmentally Controlled Condition Growing Rooms.

Natural Environmental Conditions Growing Rooms

The natural environmental condition growing rooms will be of simple construction consisting of single brick wall, with cement plaster and roof made either of RCC (1:2:4) or asbestos sheet and a false ceiling (Figure 3).

To maintain proper temperature, humidity and air supply the rooms will be air tight. The forced air circulation exhaust fan with polythene ducting will be installed on top of the door, with two vents (1½′ × 1½′ size each) for exhaust on back wall, 2 feet above the floor. Racks for growing mushroom will be made of either wood or steel. Many growers in the northern plains of India, are growing mushrooms in sheds made entirely of '*Sarkanda*' grass on bamboo racks. There is no control of temperature humidity and aeration and is entirely dependent on natural conditions.

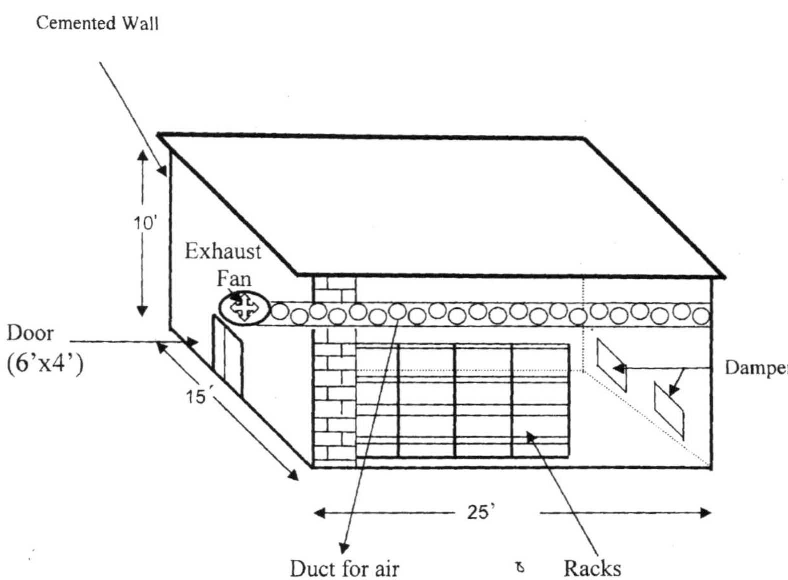

Figure 3: Natural Environmental Conditions Growing Rooms

Environmentally Controlled Conditions Growing Rooms

The environment controlled cropping rooms will be built like hermetically sealed chambers where the air movement will be controlled with mechanically control system. An ideal and profitable growing rooms of the dimensions of 45′ × 20′x 12′(h) will be sufficient to take 10-12 tonnes of ready compost at a time. The floor are generally made from reinforced concrete finished without joints and should be very smooth. A well insulated floor can be constructed by first spreading a layer of sand 6–8" thick and on top of this is poured a 2" thick layer of concrete. For providing a heating system in the floor, the plastic pipe are laid first and then 3–4" concrete flooring (1:3:6) and insulated with thermocol/glass wool 2" thick. Then it is covered with PVC sheeting and finally with 2" concrete floor. The floor should slope down towards the centre corridor and has drain. Sewerage system should be well laid (Veddar, 1978). The walls of cropping rooms should be made of brick 9" thick (length wise) with fine coating of cemented plaster. The insulation sheet of 5cm thick thermocol/ glass wool/polyurethane should be then fixed on walls with hot coal tar. One inch layer of cement plaster is then applied on the top of this and given a smooth finish.

The roof of the cropping room will be made of RCC (1 : 2 : 4), 5–6" thick. The inside should be given a cement plaster finish for application of insulation as done for walls. The roof on the outside is protected by tarring it on top, followed by 4″ thick loose soil, 2" thick mud capping and finally the tiles. This will protect the roof from weathering effects of rain and will ensure longer life of insulation and prevent seepage of moisture into the room in rainy season.

The doors of cropping rooms should also be insulated properly and made of either steal or wooden frame and covered on both sides with aluminum sheets. Insulation is done with 2" chick glass wool. The doors should be equipped with round hings and should seal to the floor with a rubber strips. The cropping rooms either, Natural or Environmentally Controlled will be built five on each side (Figure 1) with common gallery or corridor in the centre. The A.C. and boiler room will be made near to the growing rooms.

The work corridor includes the entire width in front of the growing rooms and is covered by the same roof. The width of the corridor should be 15–20 feet with a height of 10–12 feet, to accommodate the installation of ventilation ducts, heating system, pipes etc. For maintaining proper temperature required in the growing rooms during spawn run heating coils in the duct or pipes should be installed in each room. For ten growing rooms with 900 sq. feet bed area in each room or steam boiler with 300 lbs/hour (1,50,000 K Cal./hour) steam production capacity should be sufficient. In order to control the temperature, the oxygen supply and removal of gaseous waste product (CO_2) it is necessary to change the air in the growing rooms several times per hour. To obtain an equal climate throughout the room, ample air circulation is essential. This quantity of ventilated and circulated air can be achieved by equipping the ventilators with on/off timers or with speed control devices or by means of damper or shut tee (Veddar, 1978). The exhaust vents are normally made on the opposite end of the entry point, in the lower part of the wall at ground level. The exhaust vents are fitted with wire net, luvers and insulated lids. The luvers allow the CO_2 laiden air to exhaust under positive pressure created by the blower inside the air handling unit.

It is also essential to filter the ventilating air to avoid the spores of certain diseases in the mushroom farm known to spread by spores by fitting filters of (3-5μ) at air velocity of 5–6.5 feet per second (Veddar, 1978).

A permanent water supply is very necessary for fitted taps on the composting yard. One separate water pipe line of ¾″ to 1" diameter for delivering clean water for spraying should be provided in room. Underground drainage line for carrying the washing from room and wash basin discharge should be properly laid before construction of the building and connected to central gallery.

Since the cropping rooms will have no windows it is completely dependent on artificial light. Therefore, in each room, fixed lighting points are installed in the ceiling of the central corridor. Tube lights must also be installed vertically and at even heights in order to every bed to receive sufficient light. Between every two rooms, there should be points, where electric motors for conveyor belts, spawning machine, tamping machine etc. can be connected with main power.

Spawn Laboratory

It is always desirable that the spawn unit should be isolated from composting yard and growing rooms etc, since it requires pollution fee, clean area for carrying out various operation required for spawn production. A total in built area of 80′ × 40′ × 12′(h) should be sufficient to house entire spawn laboratory, to handle thousand of bottles per day. The area will be divided into following works areas:

1. Laboratory for housing various equipments of daily use.
2. Inoculation room
3. Incubation room
4. Boiling and filling room
5. Autoclaving room
6. Washing and cleaning room/area
7. Cold storage
8. Office/Store room
9. Corridor

It is desirable that the entire spawn unit should be insulated with thermocol insulating material with heavy insulation of cold storage and incubation room. All the entry and exit points should be fitted with filters of 2–3μ size to avoid contamination through fungal and bacterial spores.

Post Harvest Handling Unit

For long term preservation of white button mushroom, the best methods used are canning, pickling and freeze drying etc. For carrying these operations, a minimum in-built area of 800 sq. feet is required for construction of pre-cooling chamber {10′ × 10′ × 10′(h) size}, canning hall with canning line {30′ × 20′ × 10′(h)} and can store room/packaging room {10′ × 10′ × 10′(h) size}. The post harvest handling unit should be constructed adjacent to the growing rooms or mushroom house, because freshly harvested mushrooms are highly perishable and require immediate processing.

The other infrastructure should be constructed at a specified space and depending upon their requirement.

(iii) Principles and Techniques of Compost and Composting Process

Higher plants obtain nutrients from soil and synthesize the organic matter in leaves through photosynthesis. Fungi on the other hand grow in nature on all sorts of vegetable waste and they play an important part in maintaining the natural cycle. However, fungi/ mushroom being heterotrophic draw their nutrition from substrate, popularly known as compost. Mushrooms are capable of breaking down organic material that other micro organism cannot break down *Pleurotus* and *Lentinula e.g.* grow on drying or dead wood and *Volvariella, Stropharia, Macrolepiota* and *Agaricus* grow well on compost made from straw. The substratum in which the mycelium grows and on which mushroom eventually develop is known as compost. The white button mushroom, *A. bisporus* has unusual substrate requirements compared to other commercially grown mushroom. Therefore, preparing a suitably composted substrate for growing button mushroom is vital for a successful crop (Miller, 1994).

Composting is best approached through the concept of ecosystem management. Starting with the initial ingredients, composting requires the management of various chemicals and physical factors, that select for a succession of microbial population and substrate changes. The culmination of these successional event is the development of a dominant mushroom population. Composting is brought about by a variety of organisms (bacteria, fungi actinomycetes). Among these, thermophilic fungi developed during the course of plant residues and form a selective substrate for the development of *A.bisporus* at the practical exclusion of all other composting organism (Straatsma *et al.*, 1989). These fungi, therefore play a key role in the process of composting and bring about selectivity in the medium and many of them may ultimately contribute themselves towards the nutrition of mushroom (Batterley, 1993). Good compost, therefore, plays an important role in mushroom production and perform the following functions:

1. It provides a selective medium for mushroom spawn/ culture.
2. Synthetic compost could be tailored to meet the nutritional requirements of the specified cultivar.
3. A good compost prohibits the natural heating in the composting material.
4. Composting improves the physical structure of the substrate, not otherwise suitable to grow mushrooms and also enabling the beds to be filled more compactly than would be possible, where the straw to remain long and stiff.

The organisms which perform these essential tasks demands a plentiful supply of water and oxygen; the several turns which a stack is given enable the grower;

1. To adjust the amount of water present
2. To renew the supply of oxygen
3. To mix the ingredients and straw evenly and
4. To correct greasiness by addition of gypsum to those parts of stack which appear to need it.

It is therefore, essential to compost the ingredients in order to produce a medium which will remain stable and provide the essential nutrients for growth. In mushroom growing, preparation of such a compost is most difficult process and satisfactory results are obtained only after lot of experience.

Compost and Composting Process

Compost is the resultant product formed through the process of decomposition of a raw medium tailored by adding essential nutrients for growing a specific edible fungus. The main objective of the composting is to fulfill the nutritional requirements of the mushroom crop to produce a substrate that yields a lot of mushrooms. According to Miller (1994), other goals of composting include the achievement of following:

1. A suitable bulk density *i.e.* wet weight bulk density of approximately 550-600 kg/sq. meter related to straw softening and other structural changes.
2. Modification of plant materials, so that nutrients are made readily available to the mushroom crop.
3. Biological removal of readily available initial nutrients to avoid overheating and competent growth during either phase-II or spawn run.
4. Building up of an appropriate biomass and a variety of microbial products, some of these can serve as a nutrient source for the mushroom.
5. Establishment of selectivity *i.e.* compost promotes the growth of *Agaricus bisporus* over competitor organisms.
6. Build up of compost moisture content to serve as a water reservoir for the mushroom crop.

7. Modification of nitrogen into stable organic form, making nitrogen available to *Agaricus* but not prone to further ammonification by competitors.

With the achievements of above objectives/goals, the compost is considered mature *i.e.* ready for spawn run and cropping. According to Laborde (1994a) composting is essentially an aerobic process. However, there are pockets in the Central core of the pile, where oxygen cannot penetrate and an aerobic process sets in, producing malodorous gases *viz.*, methane, hydrogen sulphide etc. polluting the atmosphere (Deritex, 1991). Turning of compost is therefore done, to avoid such anaerobic condition, by properly aerating each portion of substrate. Temperature maintenance in the compost stack is an important parameter, since it favours the growth of thermophilic organisms and also brings about certain chemical changes resulting in production of N–rich lignin-humus complex and caramelization of the carbohydrates, which are selective food for mushroom mycelium. After the production of lingo-protein, cellulose is broken down by *Agaricus* mycelium for its nutrition, resulting in the formation of fruit bodies (Wood and Fermor, 1985).

Materials Suitable for Composting

Traditionally, the compost is prepared all over the world either from horse manure or from wheat straw. The latter has been increasingly used as major inputs as horse dung is not available everywhere. In addition, all sorts of vegetable waste can be used in preparation of compost. Paddy straw, corn stalks, sugarcane waste, coconut waste and pine apple as well as mixture of these products are also used in the preparation of compost, depending upon their availability and cost. These materials should preferably be freshly harvested and less than 5-6 cm in size. The above base material act as reservoir of cellulose, hemi-cellulose and lignin which is utilized by *Agaricus* mycelium, during its growth as a carbon source. They also add bulk to the compost, proper physical structure to the substrate and desired aeration.

Addition of Additives

The purpose of addition is to start the process of composting quickly in order that the microflora in the compost pile can make use of the required nutrients mainly carbohydrates and nitrogen, which are easily degradable. The nitrogen content of straw and other similar materials varies between 0.3 to 0.5 per cent. Whereas nitrogen content of good mushroom compost at spawning should be between 2.2 to 2.3 per cent. This means that in addition to organic materials N-fertilizers of approximately 10 kg nitrogen per ton of dry straw must also be added in the form of urea, ammonium sulphate or ammonium nitrate. Horse manure with straw is the most suitable materials for compost preparation. However, with the unavailability of good quality horse manure everywhere, use of this material has been restricted to few farms only. Chicken manure has proved to be the best substitute for horse manure. For maintaining a proper C/N ratio, wet brewer's grain malt sprout etc. are used as carbohydrate source. Small quantity of animal feed such as soybean meal, cotton seed meal, sunflower seed meal, wheat and rice bran etc. are often added, with nitrogen content of these materials varying between 2 to 10 per cent.

In addition to carbon and nitrogen, cultivated mushroom also requires nutritive element like P and K, which are added in the form of murate of potash and superphosphate. Gypsum is also added to the compost to improve its structure *i.e.* to keep the compost from becoming 'greasy' and to keep neutralize the pH of the compost. Further more, gypsum serves as a calcium source for mushroom converting the oxalic acid produced for mushroom mycelium into calcium oxalate. Following factors should be kept in view, while selecting a substrate.

1. The composting period should not be longer than customary with horse manure/wheat straw substrate.
2. The material should be readily available to growers at a price competitive with that of horse manure or wheat straw.
3. The productive potential should be relatively higher than that of horse manure or wheat straw.

Formulations

As a result of prolonged research on clues to specific nutritional requirements by an organisms, a number of synthetic formulae have been proposed. Long method of composting (LMC) for growing mushroom remained in vouge in India for about a decade, in which there are greater chances for getting less yields per unit bed area, besides taking more time for preparation of compost. Short method of composting (SMC) with perfection of pasteurization technique come into being during 1974-75, is being continuously improved upon for growing mushrooms under Indian conditions. Now indoor composting is being preferred over long and short method of composting. Several formulations have since been evolved in India and being adopted variously at different locations. Before executing any formulation the nitrogen level of compost should be adjusted between 1.5 to 1.75 (on dry matter weight basis) because a large part of this element is taken up by fruit bodies. The standard long and short method of compost formulation alongwith some others are listed below. The formulation for indoor composting remain the same as is used for short method, however the methodology is different. The first step in standardizing the short method of composting involving both outdoor and indoor phases with wheat straw as base material was taken in the year 1976-1977 (Shandilya, 1976; Hayes and Shandilya, 1977). This method was initially based on pasteurization by peak heating but very soon bulk pasteurization concept was introduced (Tunney, 1980) in the country.

Formulation Developed at Mushroom Research Laboratory Chambaghat (Solan)

(I) Standard Long Method of Compost Formulation No. 1

Ingredients	*Quantity(Kg)*
Wheat straw	300.00
Wheat bran	30.00
Gypsum	30.00
CAN	9.00
Urea	3.80
Murate of Potash	3.00
Super Phosphate	3.00
Furadan 3G	183.0 g
Lindane or BHC (10 per cent)	250.0 g
Molasses	5.0

Turning Schedule: 0, +6, +10, +13, +16, +19, +22, +25 and +28th day and this quantity of ingredients is sufficient for 45-60 bags (24" x 24")

(II) Modified Long Method (HRRS-Dhaulakuan) of Compost Formulation No. II

Ingredients	*Quantity(Kg)*
Wheat straw	300.00
Wheat bran	30.00
Gypsum	30.00
CAN	12.00
Urea	4.00
Murate of Potash	3.00
Super Phosphate	3.00
Molasses	6.00
Furadan 3G	183 g
Lindane or BHC (10 per cent)	0.250
Sunflower seed cake	20.00
Cotton seed cake	5.00
Mustard straw	5.00

Schedule: Remain the same as for above formulation.

(III) Long Method of Compost Formulation No. III

Ingredients	*Quantity(Kg)*
Paddy Straw	300.00
Horse Manure	100.00
CAN	9.00
Urea	3.60
Murate of Potash	3.00
Super phosphate	2.5

Other ingredients and schedule remain the same as for formulation No. I.

Short Method of Composting (SMC)

Formula I

Ingredients	*Quantity(Kg)*
Wheat straw	1000.00
Chicken Manure	400.00
Brewers grain	72.00
Wheat bran	100.00
Urea	14.50
Gypsum	30.00

Formula II

Ingredients	*Quantity(Kg)*
Horse manure	1000.00
Wheat straw	500.00
Chicken manure	300.00
Brewer's grain	72.00
Wheat bran	100.00
Urea	7.00
Gypsum	30.00

Formula III

Ingredients	*Quantity(Kg)*
Paddy straw	1000.00
Chicken manure	400.00
Wheat bran	100.00
Molasses	40.00
Urea	6.60
Gypsum	35.00

Schedule: -4, -2, 0, +2, +6, +8 and +10 day.

Formula IV

Ingredients	*Quantity(Kg)*
Wheat straw	1000.00
Chicken manure	600.00
Cotton Seed Meal	20.00
Urea	14.50
Gypsum	30.00

Schedule: -4, -2, 0, +2, +6, +8 and +10 day.

Methods of Preparation of Compost

There are three methods of compost preparation, which are as follows:

1. Long method of composting
2. Short method of composting
3. Indoor composting

Composting Technology

The composting technology employed today stem directly from the trial and error approach adopted by growers and workers in the early decade of 20th century. In the present day composting, a variety of technique and formulation exists for compost all of which incorporates the basic principles outlined by Lambert (1941). Process of compost making remains the same for long method of composting irrespective of the formulations used. Therefore, a standard procedure evolved by Mental *et al.*, (1972) for Solan Formula–I can be used as guidelines for all formulations based on long method of composting.

Straw is made wet 48 hours before making the pile, so that dry straw absorb water. The chemical fertilizers *viz.*, 6 kg CAN, 2.4 kg urea and full quantity of super phosphate and murate of potash are mixed with the wheat bran or brewer's grain (15 kg) about 12-16 hours before stacking and covered with wet gunny sacks. The process will allow the fertilizers to dissolve. The day the fertilizers mixture is mixed with wetted straw and heaped together is considered as zero day.

0 day

Spread the wetted wheat straw in a uniform layer of about 25-30 cm and sprinkle the fertilizer mixture on the top of it. Then with the help of fork, fertilizer mixture is mixed thoroughly with the wetted straw and stacked with the help of wooden plank (a rectangular block of 6′ × 6′ × 6′). The height and the width of pile is important and should be 5′ each for maintenance of uniform temperature in the entire compost pile. The prepared pile is kept as such for 5 days and leached water from bottom is collected and sprayed again on material. Within 48-60 hour after stacking, the temperature in the heap would rise to about 70–75°C.

+6 Day (1st turning)

Since the compost is an aerobic process, therefore each portion should receive proper air and water supply. Moreover, the central portion attain maximum temperature and undergoes maximum fermentation, whereas outer portion receives more air, but less temperature. In order that each part of compost pile receives equal amounts of temperature and air, the pile is turned periodically and according to schedule. 1st turning is given on 6th and during this turning 183 g of Furadan-3G is added to the mixture. For proper turning process about one feet material is removed from top and sides of pile, mixed properly by adding water and is kept separately. Similarly, central and bottom portion are mixed and kept separately. The pile is restacked by placing the central portion at the bottom, top and sides portion in the center and bottom portion on top and sides of the pile (turning process is illustrated in Figure 4).

+10 Day (2nd turning)

The pile is broken open and turned as stated above. 5 kg molasses is added to the mixture to enhance fermentation process. The ingredients starts turning black and temperature of compost pile increases further.

+13 Day (3rd turning)

Turn the stack employing the same technique as adopted on +10 day of turning. Thirty kg of gypsum is added in the 3rd turning.

+16 Day (4th turning)

Break open the heap as was done on +10th day and remake it.

+19 Day (5th turning)

Turn the heap and remake it.

+22 Day (6th turning)

Turn the heap and remake it.

+25 Day (7th turning)

Turn the heap and remake it. Also add lindane. In case lindane is not available add, diazinon 20EC @ 30 ml in one liter and mix into the compost.

+28 Day (8th turning + filling for spawning)

Check the compost for ammonia on this day and if there is no ammonia smell, the compost is filled in the trays for spawning. In

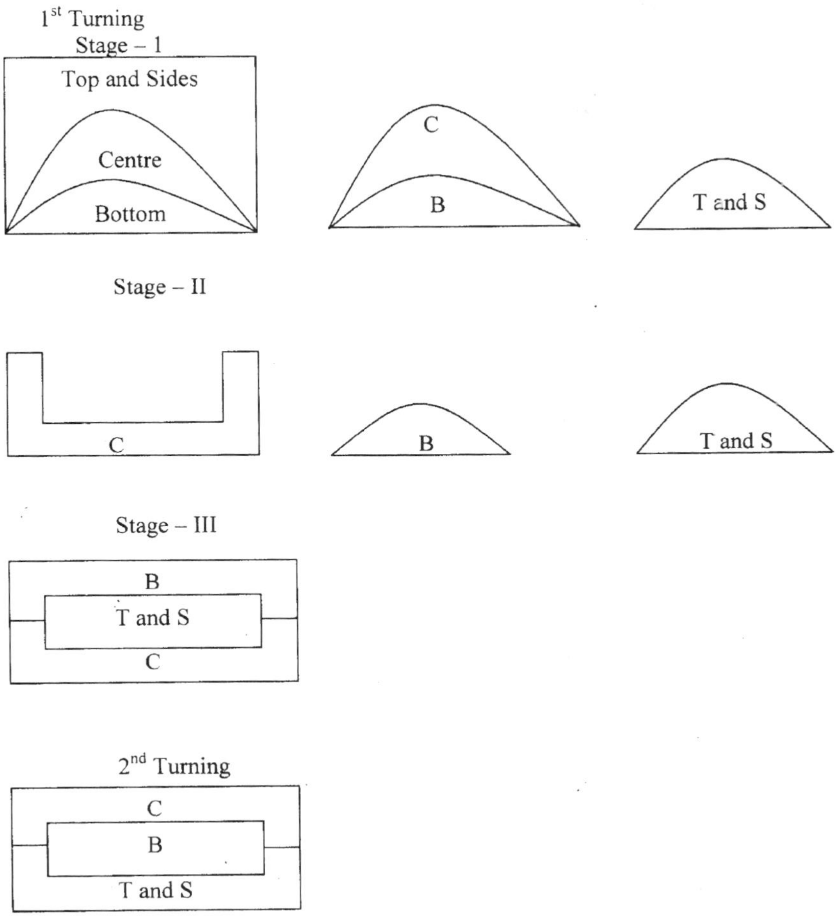

Figure 4: Turning of Compost

case ammonia, smell still persists, then turning is continued after every +3 days till ammonia smell goes off. The quantity of water added to the compost is directly related to mushroom production. If more water is added than nutrient will leach out and crop production will be poor. In order to avoid the leaching of nutrients, the run off water may be collected in a guddy pit (a small tank) and should be put back on the pile. In case the heap remain dry, on account of less water, the straw will not decompose properly and will create problems in the spawn run. A thumb rule for testing is to draw compost sample at random and press the same in the palm. The droplets of water ooze between the fingers, but should not trickle out like water from the tap. This is just the right quantity of water required in the compost. A good quality compost should have nitrogen content and pH between 2-2.3 per cent and 7.2 to 7.5 respectively for favouring best growth of *A.bisporus*. The colour of the ready compost should be dark brown and it should be free from any disease organisms, competitive moulds, insect-pest and nematodes population. The ready compost should not be sticky or greasy.

Short Method of Composting

Mushroom compost is a complex substrate offering nutrition to the mushroom crop, while discouraging the growth of potentially competing organisms. Sinden and Hauser (1950, 1953) prepared synthetic compost, consisting of wheat straw, urea and other fertilizers, where pasteurization, became integral, which was later termed as the "Short Method of Composting". The composting programme for short method of composting is absolutely different than that used for long method. It primarily consists of two phases; outdoor composting (Phase-I) of 12 days followed by pasteurization and conditioning of the compost, inside a pasteurization room/ tunnel (Phase-II) of 5-7 days or till the release of ammonia.

Pre-requisites to the Composting Process Including Machinery

Basic Infrastructure Required

1. Pasteurization room/tunnel.
2. Composting yard

Machinery and Other Equipments Required

1. Steam boiler
2. Compost Turner
3. Conveyor belt
4. Thermometers
5. Floor grills and side panels

Composting Technology

Several methods and formulae have been published by the research workers and growers from various countries. A lot of attention has been paid for evolving appropriate compost mixture with horse manure compost, based on the short method of composting. Since, horse manure is not available everywhere, particularly in India, a synthetic compost was evolved, based on wheat straw and chicken manure combination (Shandilya, 1976; Hayes and Shandilya, 1977) for commercial cultivation. The commonly used compost formulation is as follows:

Ingredients	*Quantity(Kg)*
Wheat straw	1000.00
Chicken Manure	500.00
Brewers grain	72.00
OR	
Wheat bran	100.00
OR	
Urea	14.50
Gypsum	30.00

(Now in place of Brewer's grain or wheat bran, cotton seed meal @ 20 kg per ton of wheat straw is used.)

The most important point to be taken into consideration while using a formulation, is the nitrogen percentage which should be adjusted between 1.75 to 1.80 per cent on dry weight basis, at the beginning of the process, with 75-77 per cent moisture and finish upto with 2.2 to 2.3 per cent nitrogen and 68-70 per cent moisture at spawning.

Procedure

Compost prepared by short method of composting technique has two distinct phases:

Phase-I (Outdoor composting): Completed by mixing raw materials on a platform and turning according to schedule.

Phase-II (Indoor composting): Completed by filling compost after completion of phase-I inside a room called either as a peak heating room or pasteurization chamber/tunnel. It is the conditioning of compost in closed environment.

Phase-I Process

The raw material (ingredients) is mixed together watered and periodically turned according to the following schedule:

-4 day: Straw, poultry manure and brewer's grain are consolidated by trampling to encourage uptake of moisture and an aerobic fermentation.

-2 day: The whole mass is turned and made into slightly smaller stack and more water is added.

0 day: Standard stack of 5 feet in width and height is made with the help of wooden boards. Water and urea is added to the pile as highly aerobic stack is made.

+2 day: First turning is given and restacked as above, after adding water to drier portion.

+4 day: Second turn and gypsum is added.

+6 day: Third turn.

+8 day: Fourth turn.

+10 day: Compost is ready for phase-II and can be filled into peak heating room or pasteurization tunnels manually or mechanically. The phase-I is completed as soon as composting material has become pliable, dark brown in colour, capable of holding water and odour of ammonia is sharp.

Characteristics of Compost After Phase-I and Before Filling

1. Blackish green throughout the pieces of straw gleaming wet.
2. Straw long, but can be broken with some resistance. Pieces of straw not sticking to hand.
3. Properly moist with 70-75 per cent moisture content. When squeezed firmly, drops of liquid appear between fingers.
4. Smell of ammonia in manure very much apparent, pH approximately 8.2 to 8.4.
5. Still sticky and slimy hands get dirty and wet.
6. Lightly speckled with white actinomycetes. Strongly speckled on dry spots.

7. Nitrogen content between 1.8 to 2.0 per cent and ammonia concentration around 800 ppm.

Phase-II (Pasteurization and Controlled Fermentation)

Purpose of Pasteurization

1. If the temperature during composting has been low and the compost is heterogenous, many parasites (nematodes, pathogen, flies etc.) will survive in the compost, therefore pasteurization is the best means with which these parasites can be destroyed.
2. Any excess moisture can be driven from beds for warm air will absorb more water than cold.
3. To end fermentation and to convert compost into a chemical and biological state favourable to the development of the mycelium and unfavourable to moulds.
4. It considerably reduces the time of composting.
5. Some of the food already present is made more acceptable and protein content is increased.
6. More compost per unit weight of ingredients is produced thereby increasing the yield of mushrooms.
7. Conversion of ammonia into microbial protein.

Phase-II Process

Phase-II process includes pasteurization of compost either by employing hot air live steam.

Trays evenly filled with compost are stacked in a pre-warmed room of 25′ × 15′ × 10′(h) as quickly as possible to avoid heat loss. Sufficient space is kept between the trays to allow the fresh air to move between and across the trays. Compost volume to air volume is maintained between 1:6 and 1:8. All ventilators and doors are closed. Hot air is introduced to raise the room temperature upto 35-40°C and maintained for 12 hours. By closed environment, compost temperature goes up, due to microbial activity to about 48-52°C. Fresh air is introduced slowly and gradually through controlled ventilation, so that temperature attained does not drop down for 24 hours. After pre-pasteurization stage, all door and ventilators are closed and live steam is injected to raise the air and compost temperature to 58-

59°C in order to achieve the necessary pasteurization and maintained for 4 hour with ventilation system.

Conditioning or Post Pasteurization

After achieving the necessary pasteurization, live steam introduction is stopped and dry heat applied only to maintain air temperature upto 35-40°C and compost temperature to a level of about 52-56°C for 48-56 hours or till there is no smell of ammonia. Post pasteurization and conditioning requires achieving maximum population of thermophilic flora as some of these are lost during pasteurization. Post conditioning again regenerate the lost thermophilic organism during pasteurization. Both pasteurization and conditioning make the compost selective for the growth of white button mushroom. When the compost is free from ammonia full fresh air is introduced to bring the compost temperature to the desired level for spawning which is 25°C in general.

Bulk Pasteurization

Bulk pasteurization is useful for pasteurization of compost in large quantity. The advantages of this system are very impressive; the main being the saving of time and labour. The whole process of fermentation, pasteurization and incubation can be done in 21-24 days depending upon the initial stage of the manure. According to Tunney (1978) the bulk method originated in Italy and was brought to perfection in Holland. The Dutch claim that the method is more rational than the traditional method of peak heating and mycelium growth of shelf beds or trays. Other important advantages are;

1. The odour problems are not so great.
2. More compost per square yard can be prepared in tunnel than in open and pasteurization is faster than conventional systems.
3. Shelf growers get at least one and half crops more per year.
4. That, a lot is saved on handling equipments.
5. During pasteurization and incubation difference in temperatures are only 2-3°F.

Compost treatment in bulk requires little energy and different processes are easy to control and manage. During peak heating in bulk, the required air (oxygen) can be brought where necessary, namely into the layer of compost. With the traditional peak heating method, the exchange of metabolic products and supply of oxygen takes place by means of diffusion, rather greater difference between the inside and outside of the layer of the compost may then be expected. Whereas in bulk method, difference in temperature between compost and air, will be considerably smaller and little heat has to be supplied, because the compost has the tendency to heat up of its own accord. Moreover, compost is filled directly on the grated floor without using wooden trays. Bulk chambers/tunnels of the size of 45′ × 8′x 13′(h) are sufficient for pasteurization of 25 tons of compost at a time. The compost is filled in the tunnel upto a height of 6 to 6.7′ in such a way that one square meter of space, occupies approx 900-1000 kg of compost. Several temperature sensors are placed at different points of the tunnel to measure the temperature. Immediately after filling, all the doors are closed and the blower is switched on to bring the compost at uniform temperature of 45°C. Compost temperature starts rising at the rate of 1-2°C per hour and reaches 58-59°C in next 12-13 hours. Some quantity of live steam is introduced at this stage for pasteurization or kill, which is accomplished in four hours at this temperature. It will eliminate harmful insect nematodes and competitors moulds/fungus from the compost and at the same time with preserve the nutrient in the composts, which can be properly utilized by *A. bisporus* mycelium. After pasteurization, the temperature is usually maintained at 45-55°C or till there is no detectable smell of ammonia. After conditioning, the temperature of compost is gradually brought down to 25°C by introduction of fresh air in the tunnel. The compost now will become soft and pliable with dark brown colour. It will be over layed with actinomyceted straw. Compost does not remain sticky or greasy and will smell sweetish. There will be excessive development of *Humicola* which will cover the straw with whitish aerial mycelium.

Characteristics of a Good Compost

1. The colour of the compost should be dark brown with straw dull and profuse firefangs (actinomycetes).
2. The straw should break easily, but still resilient.
3. Moisture content of the compost should be between 68 to 70 per cent. When squeezed firmly very little liquid from manure will appear between the fingers.

4. There should not be any smell of ammonia and its pH should be between 7.2 to 7.5.
5. Nitrogen content of the compost should be between 2.2 to 2.3 per cent at the time of spawning.
6. The compost should be free from nematodes population.

Important Hints

1. The heap or pile should not be too high or too wide during composting process, this will not allow proper diffusion of air to the inner part of the stack. An ideal stack should measure 1.5 m in height and width.
2. Do not over water the compost during turnings as this results in lack of oxygen in the pile. A layman's test for correct watering is that water does not run off more than 30-40cm from the pile base. Very compact stack inhibits air entry into the central part of the heap and will result in an aerobic fermentation. Moderately loose filled stack is preferred.
3. While deciding about the formulation of composting, it is always better to test the ingredients for nitrogen which should be 1.75 to 1.8 per cent on dry weight basis with 70 to 72 per cent moisture content and end up with 2.2 to 2.3 per cent nitrogen and 68-70 per cent moisture at spawning. Growers should always take care in adjusting the percentage of nitrogen and all the ingredients used in the composting should be checked accordingly.

Nitrogen content may be calculated on dry weight basis as under:

Ingredient	*Fresh Weight (kg)*	*Percentage of Water*	*Dry Weight (kg)*	*Percentage of Nitrogen*	*Weight of Nitrogen (kg)*
Wheat straw	1000.0	10	900.0	0.4	3.60
Chicken manure	400.0	20	320.0	2.6	8.30
Brewer's grain	72.0	78	15.8	5.0	0.79
Urea	14.5	0	14.5	46.0	6.67
Gypsum	30.0	0	30.0	0	0

Total dry weight	1280.30
Total weight of nitrogen	19.36
Percentage of nitrogen in dry matter	1.51

The natural product varies in moisture and nitrogen content from place to place and sample tc sample, which means that the supplies should be analysed frequently on dry weight basis.

4. Moisture content of compost can be determined by weighing overnight oven dried 100 g of compost and can be calculated by following formula:

$$\text{Moisture percentage} = \frac{\text{Wet wt.of compost} - \text{Dry wt. of compost}}{\text{Wet weight of compost}} \times 100$$

Moisture level of the compost can be adjusted by adding more water in case it is less or by evaporation if it is more.

Indoor Composting

In the past one decade, the most important change in mushroom compost preparation has been the movement to carry out composting under environmentally controlled and enclosed condition since compost prepared either by long method or short method of composting involves traditional outdoor composting which emits foul gases and smell. This necessitated the movement towards enclosed composting with following objectives.

1. Reduction in foul odors and overall environmental effect.
2. Improving the consistency of compost quality.
3. Reduction of space requirement.
4. Improvement in material handling.
5. Reduction in raw material losses during composting.

Therefore, need was felt to control the composting process in such a manner so that there is least possibilities of environmental pollution and at the same time to monitor the whole composting process/procedure specially aeration and temperature in such a fashion that an end product most suitable for growth of mushroom mycelium is obtained with a clean procedure on shortest possible time (Harper *et al.*, 1992). Work on production of such substrate stated in 1967 at France, where whole composting process was carried out indoor in specially built tunnels and hence the name indoor composting (Laborde, 1992). Since such compost is prepared under total environmental controlled conditions, sometimes, it is also called as environmentally controlled composting (ECC) (Gulliver *et al.*, 1990) or rapid indoor composting (REC) or aerated rapid composting (ARC) Laborde, 1993).

In India, such type of facilities has been created at the National Research Centre for Mushroom, Chambaghat, and work for its further improvement is in progress. Several species of thermophilic fungi have been short listed which assist appreciably in the composting process (Vijay, *et al.*, 1997). In brief the advantages of indoor composting are:

Advantages

1. There is no requirement of costly machines.
2. The requirement of composting yard is also reduced to one third.
3. There is almost negligible or very less emission of foul gases.
4. The cost of production is greatly reduced due to less number of labour.
5. Duration of composting period is greatly reduced.
6. Higher yields per unit are have been recorded.

Types of Indoor Composting

There are two types of method for preparing compost by indoor composting technique viz;

1. INRA-Method.
2. Anglo-Dutch Method.

INRA-Method

This is also called as double phase high temperature process. In this method, the entire compost mass is subjected to two different temperature ranges one higher temperature range, which is carried out at 80°C and a low temperature phase which is carried out at 45-50°C. At higher temperature, partial sterilization of compost mass occurs resulting in high production of ammonia. Bulk density of the compost mass also increases and compost becomes dark brown to black in colour due to chemical reaction.

Since most of the useful thermophilic organism are eliminated at 80°C. In this phase inoculation is done, whereby mixed population of thermophilic organisms is added to the compost mass. The temperature is further raised to 58-59°C for 3-4 hours for pasteurization which is followed by conditioning at 46-50°C for 5-7 days. Total duration of composting is 13-14 days with dry matter loss of 25 per cent.

The procedure consists of pre soaking, turning and mixing of ingredients for 2-3 days including chicken manure and addition of ammonium sulphate and filling the compost mass in tunnel. The temperature is maintained at 80°C by manipulating ventilation, through plenum from both the doors for 3 days. The tunnel is emptied and ingredients turned and the entire compost mass is filled in another tunnel with temperature 45-57°C. The temperature of tunnel is increased to 56-57°C for 3-4 hours and again reduced to 45-50°C for 5-6 days. Now the compost is cooled down to 25°C for spawning.

Anglo-Dutch Method

This is a single stage process wherein compost is kept in the tunnel at 45-50°C so that abundant thermophilic flora develops. Total length of composting cycle is 10-12 days involving the following steps.

Mix and soak the ingredients, make a heap/pile on aerated floor at about 6' in width and turn daily. Keep the compost outdoor for 4-5 days. Fill the compost mass in the tunnel fitted with nets. Maintain the low temperature around 45-50°C. Pasteurize at 56-57°C for 4-5 hours. Total duration in tunnel is 6 days and total duration of composting is 10-12 days.

Quality of incoming raw material is very important while preparing the compost by indoor method. Raw materials requirements are:

Ingredients	*Quantity(Kg)*
Wheat straw	1000
Poultry manure	600
Gypsum	30
Ammonium sulphate	
or	
Calcium Ammonium nitrate	15
Cotton seed cake	20

Step-wise Schedule of Composting

INRA–Method

Pre-soaking of ingredients

↓

Turning and properly mixing of ingredients for 2-3 days after adding chicken manure

↓

Addition of either Ammonium sulphate or CAN or Urea

↓

Filling of the compost mass in tunnel

↓

Providing ventilation through plenum and from both doors for 3 days and maintaining temperature of 40°C at bottom and 80°C at centre of pile

↓

Take out compost, remix again by adding water and fill in another tunnel

↓

Keep the temperature of the pile at 45-50°C, raise to 57°C for 4 hours and again keep at 45-50°C for 5-6 days

↓

Cool the compost and spawn at 25°C

Anglo-Dutch Method

Mix and soak the ingredients

↓

Make a uniform heap of about 6' width on aerated floor

↓

Turn daily and keep compost outdoor for 4-5 days

↓

Fill the compost in tunnel fitted with nets

↓

Maintain temperature around 45-50° C

↓

Pasteurize at 56-57°C for 4-5 hours and keep in tunnel for 6 days

↓

Cool the compost and spawn at 25°C

(iv) Spawn-Principles and Techniques of Spawn Production and Spawning

Like seed of crop plants mushroom cultivation start from spawn, the vegetative seed material. In nature, mushroom spores helps in survival of a species from one generation to next. These spores are extremely small microscopic propagules and therefore difficult to handle as seed. Spores need sufficient time and ambient condition to germinate, in order to avoid competitor fungi from exhausting the available substrate. Early records show that during the 17th century, when the cultivation of mushroom was first started, the mushroom growers used to plant horse manure beds with horse droppings containing living mycelium, which they called spawn. Therefore, spawn comprises mycelium of the mushroom and a supporting medium, which provides nutrition to the fungus during its growth. Culpeper (1652) described spawn as mycelium of mushroom, which during that time was gathered from wild (Flegg *et al.*, 1985).

Kligman (1950) defined spawn as merely vegetative mycelium from a selected mushroom grown on a convenient medium. Sinden (1972) stated that the vegetative growth is called the mycelium and the material used to plant the compost is called spawn, therefore, spawn consists of mycelium and supporting medium. According to San Antonio (1984), the word 'spawn' is derived from an old French verb 'espandre' to expand. In mushroom industry, both the vegetative body (mycelium) of the mushroom which consists of a mass of fine white threads (hyphae) and the planting material, which consists of mycelium and its substrate is called 'spawn'.

Modern spawn making began in France with Constentin and Matruchot's announcement that they have produced mushroom in compost using a pure spawn culture derived from germinated *Agaricus* basidiospores. Duggar (1905) gave the method of preparation of pure culture through tissue culture. About 1918, bottled culture spawn became available in U.S.A. Spawn planted by growers however was not itself a pure culture until later, when substrate used to produce spawn was sterilized (Jacob, 1922). The next important advance was the 'Grain Spawn' patented by Sinden in 1932. This spawn was made from whole or crushed wheat, barley or rye grains and in 1937 a patented improvement was the addition of a soluble calcium salt to the medium. In 1933, Rettew patented his method of making spawn from tobacco stems to which had been added peat humus, potato starch and lime. In 1941, Lescarboura patented a spawn based on root vegetable pulp such as beet pulp. Spawn prepared on manure substrate was common till 1940-50 (Ware and Glasscock, 1946). However, grain spawn is now prepared and used worldwide. Use of grains as substrate for spawn has facilitated the process of making spawn with greater ease and reliability. Grain spawn with larger number of loci helps to develop greater number of contact points with the substrate than the compost spawn, resulting in uniform and quick spawn run.

Before the advent of grain spawn different kinds of spawn were natural or vigin spawn (from pasteures and meadows), flake spawn (breaking of beds through which mushroom mycelium has run), mill track spawn (brick dried and made from the mixture of horse dung, cowdung and loam soil) and powder spawn.

A wide variety of various carbohydrates materials were found suitable for spawn preparation, when adjusted to a slightly acidic reaction and supplied with organic nitrogen and the usual mineral salts. This established the fact that the spawn could be prepared not necessarily on a natural substrate, but on many other organic substrate also, which could provide suitable food material for the fungus. Hu and Lin (1972) used shell powder, starch compost powder and grain hull powder for making granular spawn in Taiwan, whereas Xiang (1991) developed a new granular structure medium from saw dust or several straw powder, wheat and soybean powder for spawn production.

There was a lot of discussion and suggestion regarding the use of different spawn *viz.*, grain spawn, manure spawn and tobacco spawn. Those who prefer grain spawn, claim that there is considerable saving of time in planting, it run more rapidly and mushroom frequently appear a week earlier than manure spawn. The tobacco spawn enthusiasts on the other hand, stress the porous nature of stems, which permits aeration and promotes maximum growth, the presence of chemical substances, which nourish the mushroom, the residues of nicotine, which repels insect-pests, whereas grain spawn attracts mice and rats. Now a days only grain spawn is being used for seeding of compost by mushroom growers, throughout the world. In most of the European countries rye grains are preferred over other grains, whereas in India, wheat, jowar and bajra are commonly used for spawn production.

Modern Manufacture of Spawn

Use of cereal grains as substrate for spawn production was an important advance in making commercial spawn by Sinden in 1932. Stoller (1962) analyzed a number of practical problems in modern spawn making, evaluated patents on spawn and spawn making. Maul *et al.* (1980) described a system that can be used to make spawn entirely by bulk operation. Lots of improvements have been perfected in various steps involved in spawn production technology. The milk or glucose bottles used for commercial spawn preparation have now been completely replaced with heat resistant polypropylene bags. Modern spawn laboratories now use polypropylene microfilm windows for aeration. According to Suman (1990), the process of spawn production technology can be divided into following steps.

1. Laying out of a well equipped spawn laboratory.
2. Preparation, maintenance and conservation of pure culture.
3. Preparation of mother/stock culture from pure culture.
4. Preparation of spawn or planting material.
5. Storage and transportation of spawn.

Laying Out of a Well Equipped Spawn Laboratory

The isolation, purification and maintenance of mushroom cultures require technical expertise and aseptic high-tech laboratory facilities. Therefore, for preparation of pure culture spawn, a well equipped spawn laboratory is the basic requirement. The physical size and layout of spawn laboratory depends upon the extent and range of activities it is proposed to undertake. A built area of 80′ × 50′ × 12′(h) can handle thousands of spawn bags/bottles per day. The built area can be divided into different works area, depending upon the requirements. For example, there should be a room measuring 20′ × 15′ with in the building to serve as inoculation room, another two rooms measuring 25′x 15′ for incubation purposes. Proper space for store/office room, boiling/autoclaving room, washing/cleaning room, mixing/filling room and cold storage room have to be provided in the built area. Culture bank room for maintaining and conserving various cultures, should also be built in the building. Similarly, laboratory of the size of 30′ × 25′ will be required for keeping various equipments of routinal use and for preparation of pure culture media etc. The entire spawn laboratory should preferably be insulated with 2″ thick insulating material, with heavy insulation of cold storage, culture bank and incubation room. The spawn laboratory should be well connected with a permanent clean water supply, proper drainage system and uninterrupted water supply. A layout plan for a modern spawn laboratory is given in Figure 5.

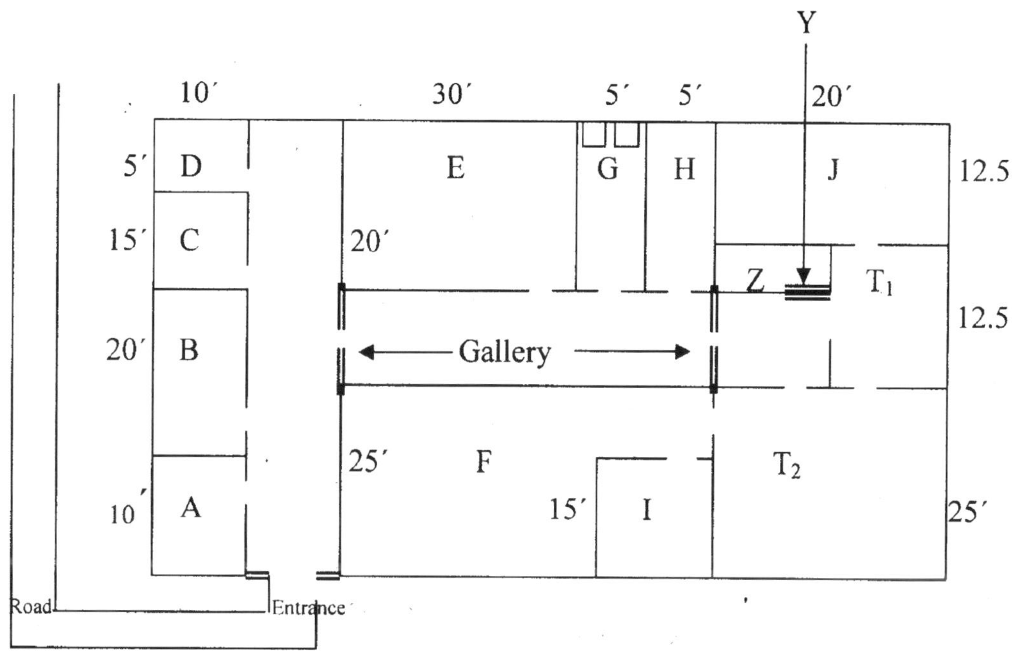

A: Office; B: Pure culture bank; C: Store room; D: Toilets; E: Cold store; F: Laboratory; G: Washing room; H: Filling room; I : Boiling and autoclave room; J : Inoculation room; Z: Sterile room; T_1T_2: Temp.controlled rooms; Y: Filter

Figure 5: General Layout of a Modern Spawn Laboratory

Special attention should be given to the design of the inoculation room. This is the place for sterile work, so the room should have an aseptic environment. An ante-room at the entrance can act as a protective buffer between the inoculation room and the outside environment. When the ante-room is opened, the sterile room should be closed.

Equipments and Chemicals

The following major equipments are required for commercial spawn production:

1. One or two autoclave either horizontal or vertical (100 kg capacity).
2. One or two incubator (6′ × 3′ × 3′)
3. One refrigerator (300 lt. capacity).
4. Boiling and soaking pans 2 nos. (50 lt. capacity).
5. 2 grains storage bins (15 quintals capacity each).
6. One laminar air flow cabinet (6′ × 4′ size).
7. pH meter.
8. Steam boiler (300 lbs capacity).
9. Racks for incubation room.
10. 2.5 ton A/C for incubation room (with heating and cooling system).
11. Working tables and chairs/stools.
12. Trolleys for transporting bags.

The following minor equipments and consumables are also required for commercial spawn production.

Spirit/UV lamps, inoculating needles, scalpels, wheat, non-absorbent cotton, polypropylene bags/bottles, match box, rubber bands, felt-tipped marker pens, test tube, labels, paraffin, aluminum foil and big size sieve mounted on iron frame.

The Basic Chemicals

1. Dextrose
2. Agar-agar powder
3. Alcohol/spirit

4. Mineral oils
5. Chlorox lactic acid, lactophenol
6. Formaldehyde
7. Calcium carbonate
8. Calcium sulphate
9. Yeast extract
10. Peptone
11. Malt extract

Preparation, Maintenance and Conservation of Pure Cultures

There are two ways of raising pure cultures.

1. Tissue culture raised from mushroom tissues.
2. Spore culture raised from single or multispores.

Preparation of Culture Media

Culture media are required for raising of pure culture. These culture media are used as substrate for isolation, multiplication, maintenance and preservation of mushroom cultures. The most commonly used culture media are follows:

Malt Extract Agar

Malt extract: 20g

Agar-agar powder: 20g

Water: 1 liter

pH: 6.5–7.0

20g of malt extract is boiled in one lt. of distilled water for 2 to 3 minutes. 20g agar is added to this and pH adjusted to 6.5. This material is then filled in glass test tubes @ 6ml per test tube, plugged with non absorbent cotton and sterilized at 15 lbs pressure for 20 minutes. After cooling down, slants are prepared. This medium is generally used for raising pure culture of white button mushroom *Agaricus bisporus*.

Potato Dextrose Agar

Peeled potato: 200g

Dextrose: 20g

Agar-agar powder: 20g

pH: 6.5–7.0

Potato are peeled, washed, cut into small size and boiled in water for 20-25 minutes. The potato extract is filtered through the muslin cloth and volume of this filtrate is raised to one liter. To this is added 20g each of dextrose and agar-agar powder, with pH adjusted to 6.5 to 7.0. After boiling for 4-5 minutes, 8 ml of this material is filled in each test tube, plugged with non-absorbent cotton and sterilized at 15 lbs pressure per square inch for 20 minutes. After cooling down stants are prepared. This medium is generally used for raising pure culture of dhingri *Pleurotus sajor-caju*.

Tissue Culture Raised from Mushroom Tissues

A large sized healthy mushroom with membrane intact on the lower side is taken from the cropping trays/bags. Lower portion of stem is cut with sharp knife, washed with running tap water to remove traces of soil particles adhering to the surface of the cap and stem. This is then dipped in 0.1 per cent solution of mercuric chloride for 30-60 seconds for surface disinfection.

A small piece of tissue from the junction of the cap and stem, is taken out with a sterilized inoculating needle and the same is transferred aseptically on 2 per cent malt extract agar medium in already prepared culture tubes. The inoculated tubes are incubated at 24±1°C for 10-15 days till such time, the surface of the medium is fully covered with the mycelial growth.

Spore Culture Raised from Single or Multispores

In the raising of mono or multispore cultures, spores are collected from a large sized healthy mushroom with veil intact under sterile conditions. The sporophore is mounted on a wire stand over a petridish under sterilized glass beaker, after the surface disinfection process as described earlier. The spores are discharged on to the petridish, on the opening of the veil in 1-2 days by itself in a thick mass. The spore print is stored under sterile conditions in a refrigerator for future use. These spores are used for direct inoculation of wheat agar medium or Lambert's medium prepared as below:

Wheat Agar Medium

Thirty two gms of wheat grains are boiled with 1 liter of distilled water for about 2 hour and filtered after 24 h. Twenty gms agar is then added to 1 liter of filtrate. The pH of the medium is adjusted at 6.5 and sterilized at 15 lb psi. for 20 minutes.

Lambert's Medium

Ten gms glucose, 0.5g $MgSo_4$ $7H_go$, 1.9g KH_2PO_4 and 20 g agar are added to 1 liter distilled water. The remaining process remains the same as described above. Spore suspension is prepared in the sterilized distilled water. Liquid agar medium (45°C) is poured in culture tubes (9 ml) and spore suspension prepared separately in sterilized distilled water is mixed (1ml) in each test tube and slants are prepared. The slants are then incubated at 28°C for spore germination for about 2 weeks. The mycelial threads will become visible on slants surface. The single spore (monospore) cultures are prepared by employing the following process.

One ml of spore suspension containing about 15-20 spores is transferred with the help of sterilized pipette in a 9cm sterilized petridish. Sixteen ml of agar medium is poured in each petridish and stirred clockwise slowly to distribute the spores evenly on the petridish surface. Turn the plated petridishes upside down and place some grain spawn in the lid. About 10 days after spore sowing, the first mycelium of germinated spores will become visible. The lid containing spawn is exchanged with a sterilized lid at this stage. The monospore is cut out with very little medium and transferred to the fresh sterilized medium filled in tubes. It is also possible to isolate germinating spores with a well established technique. The screening of single spore culture is further done after 2-3 weeks for fluffy, oppressed or strandy types. The latter is the best for multiplication. It must be remembered that all the monosporous cultures of mushrooms of heterothallic nature (*Agaricus bitorquis, Lentinula edodes*) are sterile. In secondary homothallic fungus (*Agaricus bisporus*), 30 per cent of the monosporous are sterile.

As far as tissue and multisporous cultures are concerned, they tend to reproduce the same type of parental characteristics from which these have been raised whereas there is large variation in the characteristics of monosporous cultures and selection is therefore a pre-requisite.

Maintenance and Conservation of Pure Culture

Most mushroom pure cultures properly labeled or coded are maintained on a suitable medium preferably compost extract agar medium in glass test tubes and kept at 4°C in refrigerator. They are periodically sub-cultured after 45 to 60 days to maintain vigor and viability of cultures. The tubes are generally plugged with cotton wool, which allow for satisfactory gas exchange. Mites are sometimes a nuisance in cultures not preserved in oil or not maintained in liquid nitrogen storage. Mites move from one culture to another carrying various fungal spores and bacteria with them. One way to keep the mites out of the cultures, is to cover the cotton wool plug with cigarette paper which has pores that allow air to pass, but which are too small for mites to pass through.

Another way to prevent mite infestation and to inhibit agar desiccation is to add previously sterilized mineral oil to the fluffy grown culture. The mineral oil should be of technical grade with a specific gravity of 0.865–0.890. It should be pre-sterilized in the autoclave for 121°C for at least half an hour or for 20 minutes for two consecutive days. The cooled oil should be added upto a depth of 1 cm above the top of agar slants of fully grown culture. In this way the culture can be kept upto 3 years (Smith and Kolkowrki, 1996). Retrieval is done by removing a small piece of the fungal colony, grain spawn or disc with a sterile needle, hook or loop, draining off as much oil as possible and streaking the inoculum onto agar in plates or tubes. Boeswinkel (1976) reported successful water storage of all mushroom cultures except *Volvariella volvacea* upto 7 years. The cultures are grown on a suitable culture medium and after full growth, 4-5 bits of 5mm diameter are transferred aseptically to pre-cooled and sterilized Mc Cartney bottles, containing simple water and the lids tightly screwed down. These are then stored at room temperature.

Lyophilization, also known as freeze drying is a method of choice for long-term preservation of spore bearing fungi. In freeze drying, mycelia are frozen and water is removed by sublimation. The drying of spores and or mycelia is accomplished by freezing under reduced pressure in a vacuum. Most commonly used suspensing media for freeze drying are skimmed milk (10 per cent) or trypticase soyabroth (0.75g) with sucrose (10g) and Bovine serum albumin (5.0g) in 100ml distilled water which are used with equal volume of culture suspension.

The cryopreservation of microorganisms at ultra low temperature (-196°C in liquid nitrogen) is at present regarded as the best method of culture preservation (Kirsop and Doyle, 1991). At -196°C dormancy is induced, during which the organism does not undergo any change phenotypically or genotypically, provided adequate care is taken during freezing and thawing. This method can be applied to both sporulating and non-sporulating cultures. Elliott (1986) found 10 per cent acqueous glycerol solution used as cryoprotectant was good in preserving cultures of *Agaricus* species, *Polyporus* species, *Volvariella bombycina* but not of *V. volvacea*. Mushroom cultures can be preserved by this method for more than 10 years without loosing viability. Optimization of the technique for individual strain has enabled the preservation of organisms that have previously failed (Morris *et al.* 1988). Cultures can also be preserved at -130°C in mechanical freezer upto 5 years.

In addition to above method, a new granular structure culture medium developed by Xiang (1991) can be used as an economic substitute for the traditional grain spawn medium. The substrate is prepared by mixing straw or sawdust powder (72 per cent), wheat powder (20 per cent), soyabean powder (5.5 per cent), complex additives (2 per cent) and adhesive (0.5 per cent) and sterilizing in 500ml jars. The mycelial viability and economic properties of mushroom strain can be retained for at least five years, if the mycelium is preserved at 2-4°C on granular structure medium.

Preparation of Mother or Stock Culture from Pure Culture

Next step in the spawn production process is the preparation of mother or stock culture. Pure culture raised either from tissue or spore is inoculated in a suitable substrate/medium (wheat, jowar, bajra, rye or kodograins), which provides food to the mycelium. 10 kg of wheat grain are boiled in 15 liter of water for 20-25 minutes and allowed to remain soaked in hot water for another 20-25 minutes without heating. Water is then drained off and grains are put over a sieve to dry slightly. Next day, 180g gypsum ($CaSO_4 2H_2o$) and 60g lime ($CaCO_3$) are mixed with 15 kg of boiled grains. The gypsum prevents the sticking of grains together and the lime is necessary to adjust pH (6.5 to 6.7).

Properly mixed grains are filled in bottles (either glucose or milk bottles) upto 2/3rd of their capacity (approx. 250g in each) plugged with non-absorbent cotton and sterilized at 22 lbs p.s.i. for 2 hours. Sterilized bottles are taken out from autoclave, while still hot are shaken vigorously to avoid clump formation of the grains. After cooling down, these bottles are surface sterilized with 2 per cent formalin solution and shifted to inoculation room. These bottles are kept under U.V.lights for 45-60 minutes and then inoculated with pure culture. The inoculation is accomplished by placing two bits of agar medium, colonized with mycelium of pure culture, just opposite to each other in the inner side of glass bottles and incubated at 25°C. After 7 days, these are shaken properly and carefully, so that mycelial theads are broken and mixed with grains for further fast spreading of mycelium. After 20-25 days, the mycelium will cover the entire grains and stock culture will be ready for multiplication. One bottle of stock culture is sufficient for 20-25 bottles polypropylene bags for multiplication.

Preparation of Spawn or Planting Material

Substrates remain the same and spawn is also prepared by employing the same technique as for the stock culture. One bottle of stock culture is used for inoculation of 25 bags/bottles of spawn and incubated at 25°C for 15-20 days or till such time, the grains are fully impregnated with mycelium of mushroom. The ready spawn is kept in cold storage at 2-4°C.

Characteristics of Good Spawn

Quality of spawn is mainly determined by the biological value of the strain used and the technology involved at different steps of spawn production. The spawn should be fast growing in the compost, early cropping after casing, high yielding and should produce better quality mushrooms. Mushroom growers cannot guess or predict whether the spawn which is being supplied to them is having such qualities, unless crop is grown from the spawn. Some experience based on long term research are given below as guidance to spawn buyers:

1. The substrate on which spawn is made also affects production. Spawn prepared on jowar (*Sorghum biocolor*) or wheat grains gives the maximum yield over the spawn prepared on bajra, barley or rye grains.

2. There should be a proper coating of the mycelium around every grain used as a substrate for spawn production. No loose grains should be observed in the bottles, when these are bound together with the mycelium. The grains left over, without mycelial coating will invite contamination in the compost during spawn running period.
3. The growth of the mycelium in the spawn bottles should be silky/strandy type. It should not be cottony type growth, because there is likelihood of stroma formation on the casing layer which interferes with the air exchange and absorption of water in the casing material resulting in low yields.
4. The growth of fresh spawn is more or less white. Brown colouration develops as spawn grows. Fresh spawn gives higher yield than the old one. Spawn should not be more than one month old in any case.
5. There should not be any slimy growth in spawn bottles which is an indication of bacterial contamination.
6. There should not be any greenish or blackish spot in the spawn bottles. Such types of spots indicate that the spawn is contaminated with moulds.
7. No foul smell should be present in the spawn bags/bottles.

Storage and Transportation of Spawn

Always use fresh spawn and its storage should be avoided. However, under certain unavoidable circumstances, the spawn can be stored between 2-4°C for one month. Stored spawn, before use, should preferably be kept at room temperature for some time to revive its metabolic activity. During transportation of spawn from one place to another the temperature should not rise above 35°C in any case. Therefore, they should be properly packed in thermocol boxes containing ice cubes in standing position, without dislocating their cotton plugs. It is always advisable to carry the spawn bags/bottles during night hours or transported in refrigerated van in summer months during day time.

Spawning Technology

After composting and pasteurization it is essential that the mushroom mycelium begin to grow as soon as possible in the compost which has now become selective. If growth is not established quickly, competitive mould will develop to hamper the development of the mycelium. Fast growing mushroom mycelium has the antagonistic properties. The planting material used by the growers is called mushroom spawn and the process of 'sowing' this material is called 'spawning'. The spawning area should be thoroughly cleaned and washed with 2 per cent formalin solution. Containers, tools and hands should also be washed thoroughly before spawning. The rooms where bags/trays are to be kept after spawning, alongwith racks is thoroughly disinfected with a solution of 2 per cent formalin + 0.5 per cent bavistin at least 26-48 hours before stacking. Similarly bags/trays should also be properly sterilized. The rate of spawning is kept between 0.5 to 0.75 per cent meaning thereby that 500 to 750g spawn is needed for one quintal of compost. The process 'spawning' can be performed by the following different methods.

Methods of Spawning

Pick-up Spawning

The oldest method of spawning was that in which overgrown pieces of compost were taken from old beds and placed in new beds. This was called 'pick-up'. Manure spawn, so called 'manure rolls' was also developed at that time. From the manure rolls, walnut size pieces grown through with mycelium were broken off and put into compost at regular distance close to surfaces. Another advance of note was the invention of grain spawn by Sinden in 1932. Grain spawn provided many initial growing points, it did not have to be broken into lumps, but could be distributed evenly over the surface of the bed or tray. For many years, it remained the most common method, often used in spawning. However, development did not stop at that point, which is evident from the following work.

Mixed Spawning or Thorough Spawning

'Mixed spawning' or 'thorough spawning' was developed by Sinden and Hauser laboratories in Gossau-Zurich (1950). Instead of distributing the spawn over the compost layer, a large quantity of spawn than usual is thoroughly mixed through the compost. This is the most popular method and offer the advantages of increased yields and profits because of the time saved and the fact that competitive moulds have less chance to develop in the compost.

Furthermore, turning improves the condition of the compost by thoroughly mixing the top layer, which is always somewhat dehydrated and dissipates the remainder of the anhydrous ammonia.

Shake-up Spawning

A variation of the above method was advocated by Rasmussen (1962) called as 'shake-up' spawning. In this method additional spawn is produced by the compost itself and the compost and spawn are mixed not at spawning but at later stage. In this method, the top layer of the bed is spawned with the normal quantity of spawn. After 8-10 days, when the mycelium has grown half way through the compost, the compost is thoroughly shaken and mixed. Very little mycelium is visible after this treatment, but after a few days, a stronger mycelial growth than normal is apparent and more anastomosis than usual occur. The chances of infection especially with virus are greater, when this method is used.

Super Spawning or Active Mycelium Spawning

Another method of spawning used frequently is known as super spawning or active mycelium spawning, which involves use of fully colonized compost as spawn. First a tray of one square meter of compost is spawned in the usual way with spawn. When the mycelium has grown through the bags of compost, the mixture is used to spawn 10 additional trays. This mixture is mixed thoroughly and 2-3 per cent cotton seed or soybean seed meal can be added. This method has the advantages of saving considerable time, but at the some time pre-disposing the compost to virus infections.

Surface Spawning

Where the entire quantity of spawn is spread on the surface of the compost after filling it in the trays/bags and than some quantity of compost is placed over the spawn to avoid contamination.

Layer Spawning

Layer spawning is of two types *viz.*, single layer and double layer spawning. In single layer spawning; only one layer of spawn is placed in the middle of compost whereas in two layered spawning, spawn is spread in two layers, first layer in the middle and second on the surface and covered with compost.

Spot Spawning

Compost is filled in the trays/bags and spawn is spread on the surface of compost and then inserted 2-3" deep into the compost with the help of thumb and covered with compost.

Optimal Quantity of Spawn

Generally 5 kg of spawn should be used per ton of compost. However, for shorter growing cycle upto 7.5 kg spawn per ton of compost is recommended. For a room 900 sq. feet growing surface with 10-12 tons of compost 50 to 75 kg of spawn is required. After spawning and tamping, the beds are covered with newspaper sheets and then moistened. The paper should hang down over the sides of the shelf beds and layer of paper should overlap at the centre. Covering the beds with paper or very thin plastic sheets will prevent the surface of compost from drying out and assure good mycelium growth through the compost upto the surface with increase in the concentration of CO_2. The colour of the compost lighten as soon as the mycelium begins to grow from the initial growing points and the mycelium itself can be seen as a fine creamy white film with smell of mycelium prominently evident in the room. The temperature of the room should be precisely maintained at 24±1°C with relative humidity between 85 to 90 per cent. To maintain the required level of humidity, the paper on the beds must be kept moistened by spraying water regularly. A few, days after spawning, the temperature of the beds may increase greatly with the maximum point being reached on the seventh or eight day. The increase is caused by continually spreading activity of growing mycelium together with the heat that accompanies it. So air circulation in the spawn running room should be so maintained to keep the room temperature at about 23°C and bed temperature at 25-27°C and carbon dioxide level at 1 per cent. According to Hermans (1988) about 20 cubic meter fresh air is required per ton of compost per hour, during spawn run (Veddar, 1978). Depending upon the quality of compost and the type and quantity of spawn used, the beds will be overgrown practically completely with mycelium after 10-15 days and the compost is ready for casing. It has been recommended by many research workers that supplementation of fully spawned compost with soyabean meal or cotton seed meal at the rate of 1 to 1.5 kg per 100 kg compost will increase the yield of button mushroom.

(v) Casing and Crop Production of *Agaricus bisporus*

The application of casing layer to spawn run compost induces the vegetative phase (mycelia to reproduction (fruiting). Production of fruit bodies on uncased compost has been reported but quantitatively their population is sparse, thus, casing layer is an important aspect in mushroom growing. Casing is applied when the compost is fully impregnated by the mycelium. The way a casing material is made and managed after application has a considerable bearing on its performance in terms of yield.

What is Casing?

Covering the top of mushroom beds after spawn run with a layer of appropriate soil mixture is known as casing. Although different materials are capable of supporting fruit bodies, their importance for obtaining a good crop has not been sufficiently realized. Peat, which is neutralized or made alkaline with calcium carbonate is extensively used in Europe is generally regarded as being the most convenient and suitable casing medium. Its unique physical structure, chemical nature, water holding capacity and microbiological status seemingly fulfils all the requirements (Hayes, 1978).

In India, suitable peat is not available therefore, well decomposed spent compost (Mentel, 1975) and farm yard manure (2-3 years old) are being used as casing medium (Hayes and Shandilya, 1977). The best result so far have been obtained when well decomposed farm yard manure (2-3 years old) is mixed with loam soil in the ratio of 1:1 (v/v) or with well decomposed spent compost (2-3 years old), with the same ratio (Hayes and Shandilya, 1977). Shandilya (2002) have reported maximum yield of *Agaricus bisporus* (25kg/100kg compost) when farm yard manure+vermi compost leached + 3 years old spent compost was used in equal volume for already the spawned run compost. Suman (2004) reported farm yard manure (2-3 years old) + coconut coir pith in the ratio of 4:1 (v/v) as the best casing material for production of *Agaricus bisporus* in India. Various other materials used in India as casing are;

1. Loam soil + sand (4:1 v/v)
2. Two years old spent compost + sand + lime (4:1:1 v/v)
3. Paper mulch + 2 years old spent compost

In addition to these, various other substrates alone or in combination have been tried by various workers. These are cattle manure, forest litter, tree back, sugarcane bagasse, rice husk, ash, biogas slurry etc.

Function of Casing

1. Mushrooms tend to form on the surface of the compost, but they are heavy and if there is nothing to support them, they may fall over and break the delicate 'roots' through which they obtain their nourishment.
2. If casing is not done, no mushroom, or very few one's will appear on the beds. The layers of casing soil seems to be necessary to promote the factors which lead to pinhead formation, *e.g.* difference in CO_2 concentration, the presence of bacterial (*Psuedomonas* spp.), changes in micro-climate etc. The layer of casing soil forms the environment in which mycelium changes from vegetative stage to the generative phase.
3. Vegetative mycelium is encouraged to fruit when it enters a medium deficient in food; it attempts to ensure its survival by producing fruit containing spores. A suitable casing material provides this medium.
4. The surface of compost dries out very readily and it is extremely tricky to replace that moisture without killing the spawn at that level. Casing supplies water for the growth and development of fruit bodies and regulate the flow of nutrient from compost to developing pinheads. On the average, 2 liter of water is required for formation of 1 kg mushroom. The casing layer act as a water reservoir.
5. One of the major function of the casing layer is to keep the compost fully grown through with mycelium in good condition.
6. Another factor which may stimulate fruiting is sudden reduction of temperature. The casing looses moisture by evaporation and after each watering a cool layer is provided which appear to shock the warmth-loving mycelium into activity.
7. The mycelium requires a moist medium in which to ramify. The moisture evaporating from the casing is readily

replaced by watering which, so long it is applied with care, does not reach the compost and kill the mycelium.

Characteristics of a Good Casing Soil

A good casing material should possess the following characteristics:

1. Good water holding capacity and more pore space percentage.
2. Capable to release harmful gases during cropping. Its water holding capacity is such that it can be watered without sealing off the compost. It should retain its proper structure, even after it has been watered frequently. Good structure and porosity are not only necessary for ease of handling but also to allow the proper exchange of gases from the beds.
3. The soil should be neutral in reaction or slightly alkaline. The pH of the casing soil should be between 7.5 to 8.0. Peat moss is generally acidic in nature (pH 4.5). To bring the pH to the desirable level, a certain amount of lime is added to it.
4. Electrical conductivity of the casing medium should be very low. High electrical conductivity of the casing soil may be due to some ionic salts present in the casing soil. Some ions like magnesium are deleterious to mushroom mycelium.
5. The casing soil should be free from harmful microorganisms and should be almost sterile.
6. It should be properly decomposed.
7. The casing material should be deficient in nutrient. It is generally known that the composition and yield of the casing material used greatly influences the quality and yield of the mushroom. Preparing a good casing soil is an important in obtaining high yields as properly composting manure pile.

It should be clear from above points that the material best suited for casing should have very high water holding and gas exchange capacity, low organic and ionic salts and slightly alkaline pH.

Disinfection of Casing Soil

The primary aim of disinfecting the soil is to kill harmful moulds, eelworms and mites without rendering the material completely sterile. It is therefore necessary to kill those organisms either with formaldehyde or with steam.

Disinfection of Casing Soil with Formaldehyde

For proper disinfection, the casing soil should be sprayed with 1 liter commercial formaldehyde (40 per cent) per cubic meter casing soil diluted with a sufficient quantity of water. There should not be any large lumps in the casing soil to hinder penetration of formaldehyde fumes. After proper drenching, the casing is made into pile and covered with plastic sheet or tarpaulin. There should not be any traces of formaldehyde in the casing material at the time of casing as it may kill mushroom mycelium or at least suppress its growth. Shandilya *et al.*, (1976) have recommended the use of benlate and steam for treatment of casing soil, since formalin is not effective below 15°C temperature.

Disinfection of Casing Soil with Steam

Casing soil can be sterilized in casing pasteurization room specially constructed for this purposes. The casing soil should not be steamed longer than 5-6 hours at temperature of 60-65°C or for 4 hours at 70-75°C. This temperature is sufficient for killing harmful microorganisms, provided that the material is not too dry. The soil is filled in wooden trays which are stacked in the pasteurization room, with sufficient space for circulation of air. The steam enters the room through a perforated pipe placed below the grated floor so that major part of the condensed water does not penetrate the casing soil. Electronic temperature probes thermometers are placed in trays to check the temperature outside the room.

Drum Method of Steaming

The drum system is an improvised method of pasteurization, which could be made at home easily. An empty 200 liter drum, fitted with two 1500 W immersion rods at the base and a wooden stand above is used for this purpose. A perforated inner cylindrical bucket is placed in the wooden platform and water is filled upto the surface of wooden structure. Casing mixture in gunny sac is placed inside for pasteurization. A long stem thermometer is inserted through the air tight top of the drum to the centre of the casing material and switched on. When steam passes through the casing material, its temperature rises and once 65°C is achieved, it is maintained for a period of 4 hour. The casing material is then taken out from the gunny bags and allowed to be cooled down to 25°C, pH level (8.0) is adjusted and material is now ready for casing purposes.

Casing Spawned Bags

The newspaper sheet is removed from the top of the mushroom beds. The spawn run compost is slightly pressed with the help of wooden board. The pasteurized casing mixture is spread evenly on the top of the mushroom beds upto a depth of 3.5 to 4.0 cm. The cased trays are then maintained at 23±2°C for 8-10 days. To make the casing layer at level, it is essential that compost be spread evenly on the beds and tamped down firmly. After casing the room is swept and spilled soil removed. The room is then cleaned especially under the beds by spraying with water. Some quantity of casing soil is saved for use in later stages. A light spray of formaldehyde at a concentration of 0.5 per cent should be given on the bed surface to avoid contamination by moulds and other microorganism.

Maintenance of Proper Humidity, Temperature and Aeration after Casing

After casing, the moisture content of the casing soil must be raised to optimal level by spraying 5-7 liter of water per 100 kg compost, for the first three or four days. The first watering can be given immediately upon casing and during 2nd watering formaldehyde at the rate of 0.1 per cent can be added. Benlate can also be sprayed in place of formaldehyde at the rate of 125-150g per 150 liter of water for 100 sq. meter bed areas. Spraying of small quantity of water may continue for 8-10 days. During pin head formation, small quantity of water is required, therefore, room should be very rapidly ventilated and CO_2 flushed out of the room.

The temperature and ventilation should be controlled so that the mycelium will grow vigorously and quickly into the casing soil and begins to form pin heads. The optimum temperature for the mycelial growth is 25-27°C during the first few days after casing, bed temperature should therefore be maintained at the level of 22-23°C. During pinhead formation the air temperature can be maintained at 15-17°C and bed temperature at 18-19°C. During the first two flushes it is advisable to maintain the temperature between 14-18°C. The growing mushroom require much oxygen and in exchange breaths out carbon dioxide (harmful above 0.2 per cent). By watering a bed and then airing, you can provide by evaporation, the cool layer on the surface of the soil which is still believed by many to encourage the spawn to fruit. Pinhead formation is induced largely by difference in carbondioxide concentration in and just above the bed. When the percentage of CO_2 is higher (1-2 per cent) the mycelium grows well (vegetative growth). After a week or so following casing, generous ventilation should be provided to induce fructification (generative stage) that is 80 to 120 cubic meter fresh air per ton of compost per hour. The pin head in the deeper part of the casing layer often die when the CO_2 concentration is too high *i.e.* above 0.2 to 0.3 per cent. Therefore, it should be maintained between 0.1 to 0.15 per cent.

(vi) Environmental Crop Management

Successful mushroom production is directly dependent on prevailing environmental conditions in a mushroom house. After case run, the vegetative phase is almost complete and we have to turn over to those conditions more favourable for reproductive growth. Depending more or less on the methods of growing, the strain used and the thickness of casing layer, the first mushroom of the first break will be ready for picking about 18-20 days after casing. Sufficient quantity of moisture should be available during first flush. The grower has to provide sufficient water in the compost and in the casing soil therefore, the air used for ventilation should have the correct degree of humidity with proper temperature in the dropping room. A view of the crop of *A. bisporus* is given in Plate 1.

When production of mushroom start, it continue regularly for 3 to 4 days after that production decreases rapidly and stops almost completely after few days. Meanwhile, new pinheads will be developing and in a period of few days, there will be another crop of mushroom. This phenomenon is called flush or a break, with a gap of at least 6-8 days between the two normally, the first, second and third flush records more than 75 per cent of the peaks as well as the total production. Thereafter it decreases gradually.

It is important that during cropping the beds should be sprayed lightly, although the relative humidity should remain high through a continuous wetting of walls and floors. Watering schedule for second flush, should be same as these of the first however in all subsequent flushes, the requirement of water will be less. Watering of beds must be adopted according to prevailing circumstances, climate outside as well as types of casing used. For growing mushroom in shelf beds, the moisture content of compost at spawning should be 66-68 per cent, in trays upto 70 per cent and in bags around 65 per cent.

Forced air circulation is very essential during entire cropping season and when there is a heavy flushes. During fructification and growth of pinheads into full grown fruit bodies, quite an amount of heat is produced in the bed, and this heat must be removed again, primarily by evaporation of water, from the upper layer of beds. Evaporation of moisture from the surface of beds is also required for the transportation of nutrient from compost layer to the mushroom fructification, therefore, there is need to renew the air inside the cropping room at least four to five times in one hour to remove CO_2 and bring in oxygen. Measurement show that CO_2 production during the first three flushes amount to 0.06 to 0.08 g per kg compost, per hour at 16°C. Therefore, one kg mushroom on beds will produce 1g CO_2 per hour under normal growing temperature and it increases by 20 per cent for every centigrade the temperature in the bed increases.

The CO_2 concentration at cropping stage should be as much as we have in fresh air (0.02 per cent to 0.03 per cent). Thus, for a flush yielding 5 kg mushroom per 100 kg compost, with a room measuring 200 sq. meter bed areas, air needed for ventilation will be 1000 cubic meter per hour.

In a growing room, one must be clear regarding ventilation and circulation. Ventilation means letting in fresh air and removing used air from room, serves to keep the percentage of CO_2 in the room below 0.1 to 0.15 per cent, whereas circulation of air is essential to reduce large temperature difference and an increase in the percentage of CO_2 in the room below 0.1 to 0.15 per cent. On the other hand circulation of air is essential to reduce large temperature difference and an increase in the percentage of CO_2 in certain areas. For an optimal production and good quality of mushroom, an air circulation of 10-15 cubic meter of bed surface is required per hour, depending on the breaks. During first 2 flushes, continuous air changes will be required, but in later flushes, the CO_2 production is reduced, thus reducing the requirement for more frequent air changes.

Temperature of air during cropping should be maintained at 14-18°C with temperature of beds slightly higher, which is brought by opening the ventilation of the cropping room and introducing the fresh air via air handling unit. The CO_2 laden air is exhausted through back vents under positive pressure created by centrifugal fan, inside the air handling unit. In the summer during a very hot period in which the temperature in the room may increase greatly, it is recommended that ventilation be frequent, especially during night and in the cool morning and evening hours. Forced air circulation system can be installed in the cropping rooms for effective air circulation inside the growing rooms connected to a perforated polythene duct running across the room. Proper watering, maintenance of temperature and air circulation inside the growing rooms is the most important and vital operation in the entire management of crop environment for raising a healthy crop. Therefore, to obtain maximum production of button mushroom, the crop should be managed employing standard parameters such as good quality compost and spawn of high yielding mushroom strain and strict control of environmental factors, *viz.*, temperature, humidity and fresh air.

(vii) Harvesting and Marketing

The first flush starts coming up 2-3 weeks after the casing and the production continues in flushes for a period of 60 to 90 days. However, a crop should not be allowed beyond a period of 60 days for economic reasons.

In order to make room for appearance of successive flushes, it is essential to harvest the mushroom daily. The buttons are harvested by holding the cap with forefingers slightly pressed against the soil and then twisted out. This helps in preventing the disturbance to the spawn run bed and other pinheads coming up. Soil particles and mycelial strands clinging to stem are cut off with a knife. Fresh casing soil must be applied or rather filled in the holes created by removal of mushroom after harvesting.

Mushroom harvesting is a routine job and proficiency comes with daily picking. Careful handling is necessary to prevent bruising and subsequent discoloration. When diseases are present in a room, care must be taken to keep them from spreading all over the farm at harvesting. A growing room with shelf bed farm and a growing surface of about 1800 sq. feet can be harvested by four men in four hours, including some extra work for removing extra mushrooms. Before fresh mushrooms can be delivered to the consumer, they have to meet certain standards. These standards can vary greatly, not only between one country and next, but even from one region of a country to the other region. Open mushrooms will be almost unacceptable in the market.

Before packing the mushrooms are graded for commercial used. There are three common grades which have varying price in the market.

Grade–A

Comprises mushrooms which are buttons, 2.5 cm to 4 cm pileus diameter, with tight and closed surface as solid white fruit bodies. Membrane securely intact, that is, likely to remain unbroken for twenty four hours.

Grade–B

Comprises mushrooms which are cups, 4 to 7.5 cm pileus diameter and veil is partially ruptured and gills are exposed to some extents, that is, the mushroom in an open stage of maturity, the cap retaining a pronounced downward curve.

Grade–C

Comprises flat (fully opened with flat cup underside). The head diameter is more than 7.5 cm in diameter. The cap is completely expanded forming a letter T with the stack and gills are fully exposed and light in weight.

Packing

Mushrooms are highly perishable commodity. The white colour turns brown black in couple of hours with prevailing high temperature and become unfit for consumption. Therefore, mushrooms are usually packed in the following material for sale or transportation purposes.

Pager Bags

Mushrooms are usually packed in brown paper bags of 200, 250 and 500 g capacity. Quantities more than this have a tendency to loose original colour and thus reduce consumer acceptability.

Polythene Bags

Mushrooms are now packed in polythene bags properly perforated so as to exchange fresh air with the accumulated gases, which are required to be replaced, for maintaining good quality. Mushrooms are usually sold in 200, 250 and 500 g capacity polythene bags.

Insulated Ice Containers

Insulated ice containers are used for transporting over 8-10 hour transport distance. The container is usually lined internally with 4" thermocol, which is covered with the sheeting on both sides to give strength. Ice is usually placed at the bottom, with a shelf put on the top in which the mushrooms are preserved. Mushroom will remain healthy and white in colour for quite a long period.

Polypacks containing mushrooms should be stored at low temperature (<5°C) till they are marketed or consumed but not more than 2 days. Mushrooms have very short life and efforts should be made to sell them fresh, as soon as possible. After completion of harvesting, the beds are cleaned and fresh of trash. Cleaning the beds has a beneficial influence on the structure of casing soil and thus on the exchange of gases in the beds. It is not important, therefore, that some pinhead may be lost at cleaning. It is better to remove, as far as possible all impurities after each flush. When the beds are cleaned at the end of flush, the next flush will appear sooner and in a more regular way. Between breaks, beds that have been picked clean, can be sprayed with formaldehyde solution at the concentration of (0.25 to 0.30 per cent) to prevent diseases, moulds and bacterial spots. After 60 days of cropping, depending upon the growing schedule the growing rooms with compost are cooked out by live steam and then emptied. The beds are picked clean, watered superficially, the electronic thermometer are placed in the compost and live steam is used to bring the compost temperature to 70°C for 12 hours. The cooking out process should be done gradually. After this operation, empty the trays and bags, away from the mushroom house immediately. The rooms with racks should be sprayed with sodium pentachlorophenate @ 2 per cent. This treatment is designed to kill the mycelium growing in the side boards of racks, so that it cannot be passed on to the next growing cycle. After the solution has remained in the room overnight, everything is washed down again, with a quantity of water. The rooms are then ready for next growing cycle.

Chapter 7

Cultivation Technology of Specialty Mushrooms

Specialty mushroom is a term given to a group of mushrooms which are less common in a particular area or country. In other words 'specialty mushroom' includes all mushrooms other than common white button mushroom which accounted for over 32 per cent of the total production value of 6.2 million tonnes mushroom production in 2000. In India, about 80-85 per cent of the approximate 60,000 tonnes of total annual production is that of button mushroom, 15-18 per cent of *Pleurotus* spp. and the remaining 1-2 per cent of the other mushroom. Varied agro climatic conditions and availability of agricultural and industrial wastes in India, offer great opportunities for cultivating mushrooms on commercial scale in our country. More than 380 million tonnes of agricultural waste is available annually in India and about half of these residues remain unused. These residues are resistant to degradation, because they contain mainly cellulose, hemicellulose and lignin. Most of these specialty mushrooms possesses the enzyme complex which enable them to attack and degrade these industrial and agricultural byproducts, thereby resulting in highly valued food protein suitable for direct consumption. In addition to this, these specialty mushrooms also represent an attractive method of improving the nutritional quality of lignocellulosic wastes for use as an animal feed stock. These have also been traditionally used in China and Japan for medicinal and tonic purposes since time commercial. Except button mushroom, all other cultivated edible genera of fungi can be grouped under specialty mushrooms. The cultivation technology of some of them is given here under:

(i) Cultivation of *Agaricus bitorquis*: The Hot Weather Mushroom

In India, attempts were made to develop cultivation technology for white button mushroom in early sixty's and since then, this has been the only cultivated white button mushroom in most parts of the country, particularly temperate regions. Since major parts of the country experiences warm climate during most of the period in a year, cultivation of *A. bisporus* was difficult during summer months, without environmental control, resulting in heavy investment. Therefore, efforts were made to introduce those mushroom cultivars of white button mushroom which could be cultivated under such conditions without incurring heavy investment on environmental control. *Agaricus bitorquis* (Quel.) Sacc. also called as hot weather white button mushroom and a Basidiomycetous fungus, was found most suitable for cultivation under sub-tropical conditions.

A. bitorquis which possesses a number of desirable attributes in comparison to white button or temperate mushroom could easily be cultivated relatively at higher temperature ranges during summer months when cultivation of *A. bisporus* is not feasible.

Compost Preparation

Compost for cultivation of *A. bitorquis* is prepared in the same manner and technique as for *A. bisporus*. It is pasteurized and conditioned without any change as prescribed for *A. bisporus*. After peak heating, the compost is cooled to 28-30°C, which is considered as optimum temperature for mycelial growth of *A. bitorquis*. The compost should be absolutely free of ammonia and must have moisture content of about 70-72 per cent at filling.

Spawning and Spawn Run

After cooling down, spawn is mixed thoroughly in the compost and the compost is tamped with the help of a tamping machine or manually. The quantity of spawn used per unit area is the same as in the case of *A. bisporus*. Quantity generally recommended is 5 to 7.5 kg spawn per tonne of compost. After thorough spawning, the compost is tamped and covered with a plastic sheet or paper. Plastic sheet is sprayed with 4 per cent solution of formalin thrice a week. A high carbon-dioxide concentration is desirable in the compost during mycelial growth period and following requirements need be maintained.

Compost temperature	:	28-30°C (82-86°F)
Relative air humidity	:	90-95 per cent
Air circulation	:	Very slight/little, so that changes in temperature does not occur.

Mycelium should be allowed to grow in the compost for atleast 15 days and the plastic sheet is removed a day before casing to dry the upper surface of the compost.

Casing Mixture and Casing

Casing mixture of F.Y.M. (2-3 years old) and Garden soil in the ratio of 1:1 v/v, is the best casing material used in our country. After properly pasteurizing the mixture, either with live steam or chemicals, the casing is applied upto a depth of 3-4cm and levelled off without being compressed. Soon after casing, 2 lt. of formalin in 100 lt of water is sprayed over a cropping area of 100 sq. meter cropping area and doors are kept closed. Formalin vapours are removed through ventilation on the following day. A high carbon dioxide level is must for *A. bitorquis* during mycelial spread period and, therefore, doors of growing rooms or mushroom houses should be air tight. The climatic conditions are similar to those as during spawn running. The climatic conditions are similar to those, as maintained during sawn running period. The casing soil is usually colonized by mycelium after 13-15 days. Use of diazinon, melathion and endosulphan is advantageous during casing to check the population build up of flies and mites.

Pinning and Cropping

As soon as the mycelium reaches the casing surface, the compost temperature is lowered to 24-26°C through ventilation and maintained throughout the cropping period. Fresh air requirements for *A.bitorquis* are generally determined by the amount of compost per square meter bed area. Maximum quantity of air is circulated until the required compost temperature (25°C) is attained for every 100 kg compost per square meter in case of K-26 and K-32 strains. Ventilation is controlled to prevent change in the compost temperature and the flushes. Too much ventilation will result in scaled mushrooms, while too little ventilation will lead to mycelial growth on top of the casing and shall promote growth of unwanted moulds.

The following climatic conditions in cropping house are desirable:

Compost temperature	:	25±1°C till the end of crop
Relative air humidity	:	About 90 per cent from development of pin heads and throughout the cropping period.
Air circulation	:	Very little, that the temperature does not change.

Watering is essentially the same as for *A. bisporus* however, watering should not be started before the mushrooms are 1–1.5 cm in diameter. The water requirement from casing till end of the crop is generally 1–1¼ liters per kg of mushroom. It is of importance to remember that air temperature does not rise and moisture content of casing does not become too high when the mushroom flush is ready for harvest. These conditions are very favourable for occurrence of bacterial spot and dry bubble diseases. In case cut surface of stem turns brown or pin heads die, it is an indication that casing soil was too wet. To get sufficiently high yields, harvesting period of six to eight weeks is maintained. In addition, proper hygienic conditions in the cropping rooms should be maintained throughout. Many disease pathogens are likely to flourish, if dead material is allowed to remain on cropping beds. In a cropping period of six to eight weeks 4 to 5 flushes can be harvested (Plate 2).

Plate 1: A View of the Crop of *A. bisporus*

Plate 2: *Agaricus bitorquis*

Plate 3: ***Pleurotus sajor-caju***

Plate 4: ***Volvariella volvacea***

Plate 5: *Calocybe indica*

Plate 6: *Lentinula edodes*

Plate 7: *Flammulina velutipes*

Plate 8: *Agrocybe aegerita*

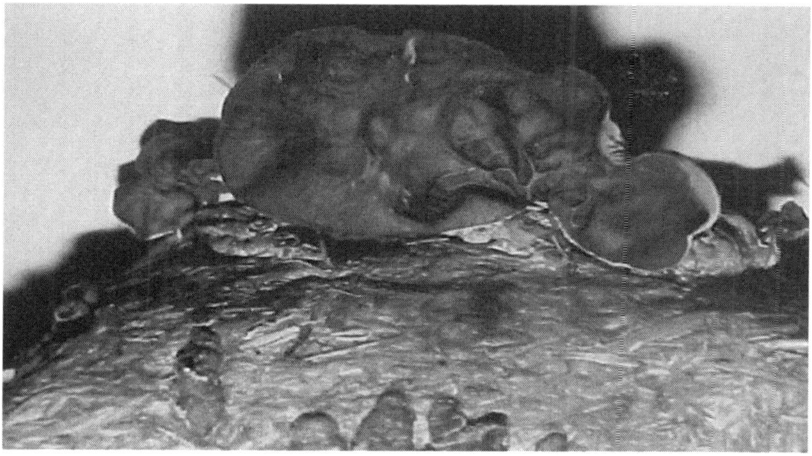

Plate 9: *Auricularia polytricha*

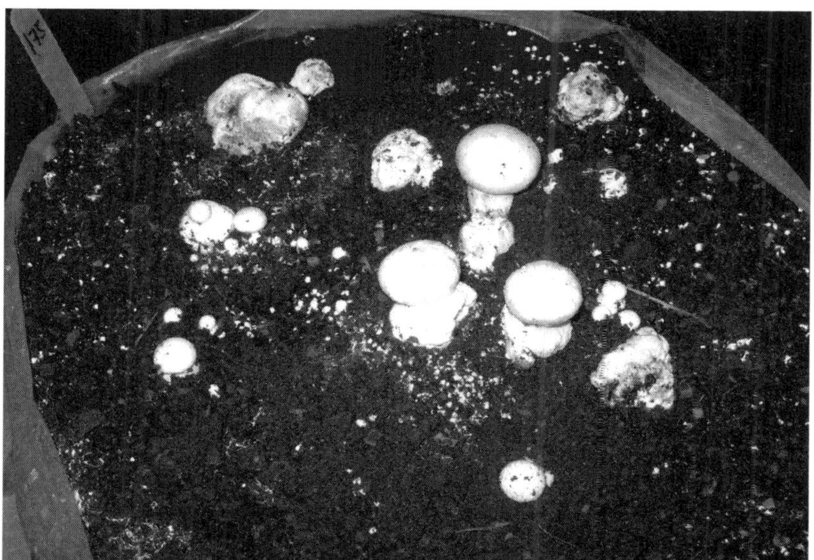

Plate 10: Symptoms of Wet Bubble

Plate 11: Symptoms of Green Mould

Plate 12: Symptoms of False Truffle

Plate 13: Symptoms of Yellow Mould

Plate 14: Symptoms of Bacterial Blotch

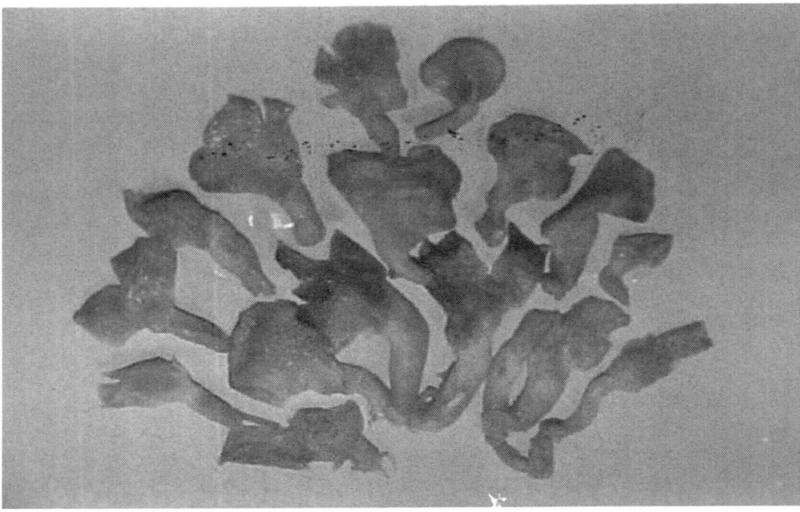

Plate 15: Symptoms of Yellow Blotch

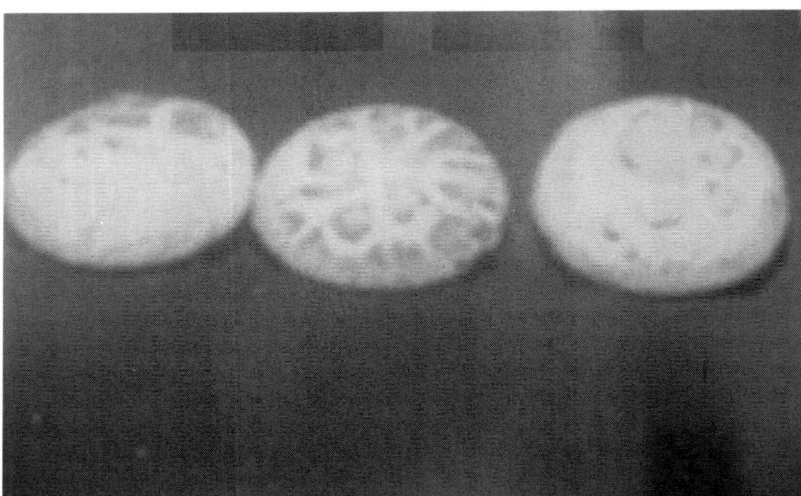

Plate 16: Mushroom cracking

Status of *A. bitorquis* in Mushroom Cultivation

One of the chief characteristics of this new species is its capability to grow at high temperature and in higher carbon dioxide concentration than its close relative *A. bisporus*. *A. bitorquis* is also resistant to viral diseases which is responsible for dreaded die back disease well known in *A.bisporus*. The fruit bodies of *A. bitorquis* are, in general, coarse and, thus, possess a better shelf life. In addition, *A. bitorquis* mushrooms are especially more resistant to pressure and bruising and are less inclined to become brown, hence earn best price in the fresh market. Being tetrasporus, *A. bitorquis* offers greater prospects of breeding new varieties. This is the only species in which different strains suitable for fresh sale and canning are available.

(ii) Cultivation of *Pleurotus* Species: The Oyster Mushroom

The oyster mushroom known as wood fungus or *'Dhingri'* belongs to the genus *Pleurotus*, has its origin from Greek word "Pleuro" which means formed laterally or in a side way position, referring to the lateral position of the stipe (stem) in relation to pileus (cap). The fruit bodies of this mushroom are distinctly shell, fan or spatula shaped with different shades of white, cream, grey, yellow pin or light brown depending upon the species. These are popular edible fungi, easy to cultivate and cost of production is quite low. Besides this, oyster mushroom can be sun dried and stored for sale at convenience on remunerative prices.

Technology for Cultivation

All commercially cultivated species of the genus *Pleurotus viz.*, *P. ostreatus* (grey oyster mushroom), *P. florida* (white oyster mushroom), *P. flabellatus* (pinkish white oyster mushroom), *P. sajor-caju*, *P. cornocopiae*, *P. sapidus*, *P. eryngii*, *P. djamore*, *P. tuber-regium*, *P. platypus*, *P. opuntiae* etc. are similar in habit and habitat. These are mainly cultivated on paddy straw, wheat straw, cotton waste and corn cobs. However, other agricultural wastes such as banana psudostems, rice husk, saw dust of broad leaf trees, coconut waste sugarcane bagasse, sunflower stalks etc. can also be used if locally available as substrate for raising commercial crops.

Method of Substrate Preparation

There are various methods to kill undesirable microorganism present in the straw to favour the growth of *Pleurotus* mycelium. The popular methods of substrate preparation are as follows:

Steam Pasteurization

The locally available substrate should be soaked in water to the extent that it contains 68-70 per cent moisture. The substrate to be used for cultivation should not be old, exposed to rain, dark brown or mouldy. The soaked material is pasteurized at 65-70°C for one to one and half hour and then allowed to cool down slowly. The entire process takes 3-5 days.

Hot Water Treatment

In case the pasteurization facilities are not available, the substrate may be treated in boiling water for similar period and allowed to cool down slowly. Many growers have been using un-pasteurized paddy straw with great success, but in such cases it should be fresh and shiny in appearance.

Chemical Treatment

Wheat or paddy straw is treated by steeping in a chemical solution of carbendazim 50 per cent WP (75 ppm) and formaldehyde (500 ppm) for a period of 16-18 hours. Most of these mould are either killed or their growth is suppressed for 25-40 days after spawning. The technique consists of steeping 10-12 kg of straw in 90 litre of water slowly in rust proof drum. In another drum 7.5g Bavistin and 125 ml. formaldehyde is dissolved in 10 liter of water and slowly poured on the already soaked wheat straw. Straw is pressed and covered with polythene sheet. After a period of over 16-18 hours, straw is taken out and excess of water is allowed to drain out. This solution is used twice for treatment and after this fresh one is prepared. However, steam and hot water treatment methods are still preferred as there are reports of phytotoxicity by using this method. The pasteurized substrate becomes ready for spawning after being cooled. Some growers prefer to supplement the substrate with protein rich substances such as wheat bran or gram dal powder but this exposes the substrate to contamination also.

Spawn, Spawning and Crop Management

Pleurotus spawn should be about 15 days old in which mycelium has formed complete coating around the grain, otherwise the naked grain will attract many competitors and pests and serves as foci for contamination. The process of spawn making is the same as in *Agaricus* species. The normal rate of spawning in a pasteurized substrate is 1.5–2.0 per cent of the wet substrate, however, it should be kept higher (2.0–2.5 per cent) in un-pasteurized material. The spawning is usually done in layers or even in through spawning care should be taken for uniform mixing of spawn with the substrate, while in layer method the spawn is mixed after each layer of 3-4 cm thickness of straw. Polythene bags (24" × 24") have been found best and cheap container for *Pleurotus* cultivation before filling the substrate in polythene bags holes of about 1 cm diameter should be made at 10-15 cm distance all over the surface for free diffusion of gases and heat generated inside. After filling the mouth of bags should be tied with sutli or just folded up.

Wooden trays, bamboo baskets or trays/shelves of wire mesh can also be used for commercial cultivation in place of polythene bags. The bags should be incubated at 25-28°C and relative humidity of more than 80 per cent. The temperature inside the bags will be 2-3°C higher than the room temperature during early stages of growth due to microbial activity. The room may be kept dark during the spawn run period. The bags should be then slightly tapered from sides and pulled from the sealed corners to expose the substrate for primordial initiation. These bags after washing and treating with 5 per cent formalin can be reused 3-4 times.

The colonized substrate blocks are kept about 15cm apart from each other on wooden benches made of 3-5cm thick wooden strips. Two to three wooden benches along the sides of the room or one above the other can be made to keep the bags. For a good commercial crop, the temperature in the cropping rooms should be 20-30°C and relative humidity above 90 percent. Fresh air should be provided by opening the windows and door for 2-3 hours or by running exhaust fan fixed at the top of the door of cropping room. During pinhead formation, that is after about 25 days of spawning, aeration should be provided after each watering, so that the surface of pinheads becomes dry. This avoids the chances of diseases/moulds attack and also the rotting of developing pinheads as otherwise pinheads are likely to turn brown to reddish brown, spongy and then die. Aeration and light also promote the initiation of pinheads. If no diffused sunlight is available to the bags, artificial light with bulb or tube may be provided for a short time *i.e.* 1–2 hours daily. Excess light changes the sporocarps to dark brown to blackish in colour. Watering should be done 1–2 times daily but the care must be taken that the initial moisture content of the substrate remains the same (*i.e.* 68–70 per cent). The first flush will be ready for harvest after about 27-31 days of spawning. Though the oyster mushroom can grow at 15-30°C but the yields at lower temperatures are lower and cropping is delayed in almost all the species except gray oyster mushroom, which need be incubated at lower temperature (12-16°C) for fruiting. The crop is obtained in intermittent flushes after a break period of 5-7 days depending upon temperature. The mycelium at the outer surface disappear and crop stop coming up in some cases at higher temperatures, when alternate drying and wetting of substrate takes place through careless management. In such cases, outer surface of the straw should be scrapped followed by watering and maintenance of proper relative humidity which restores normal cropping again. The crop continues for 60-70 days depending upon the good management practices. (Plate 3).

Harvesting of sporocarps should be done before the margins start folding in wards after which it tends to accumulate extra moisture and the desirable qualities are diminished. Slightly immature mushrooms, if harvested, can be stored for comparatively longer period. Mushrooms are conveniently packed in polythene bags with enough holes for ventilation for preventing accumulation of extra moisture, inside the packets except in rainy season. During rainy season, when there is plenty of relative humidity in the atmosphere, the shelf life is more in un-perforated bags compared to perforated bags. Direct pulling of sporocarps from the bags should be avoided as the straw pieces will come out alongwith and the smaller pinheads will be damaged. Freshly harvested mushrooms can easily be stored in a refrigerator for 4-5 days, without much loss in quality. The production can be dried in sun or in oven/ dehydrators. Care must be taken that the drying temperature should not be more than 48°C as the fat will be oxidized at higher temperature and the product will give a rancid smell. Brittleness in dried fruiting bodies is the right stage attained through drying. The dried mushroom can be stored in airtight polythene bags properly sealed in dry atmosphere. Well dried and properly stored mushrooms can stay for 1-2 years in good condition.

Cost of Production

Fresh mushroom are generally sold for Rs.25-30 per kg. The cost of production does not exceed Rs.8-10 per kg of fresh *'dhingri'* depending upon the availability of the raw material used.

Food Value

Dhingri contains fairly high quantity of ascorbic acid (Vitamin C) which helps in the treatment of scurvy. In addition to this, it is a good source of calcium and phosphorus which are essential for the formation of bones and teeth. Being low caloric and highly nutritious food, it is also recommended to counter obesity and especially useful against diabetes.

Advantages of 'Dhingri' Growing

1. Among all the cultivated mushrooms *Pleurotus* has maximum number of commercially cultivated species suitable for round the year cultivation. Various species of *Pleurotus* can grow at wider range of temperature and do not require too much exacting conditions as for *Agaricus bisporus*.
2. It grows well on fermented and non-fermented substrate and casing is not required. Substrate preparation for oyster mushroom is very simple.
3. It can tolerate higher concentrations of carbon dioxide.
4. It has nitrogen fixing ability and enriches the substrate, which can be further used as fertilizer.
5. The productivity is very high as compared to all other cultivated mushrooms. It is the only mushroom which yields equal to the inputs. One kg dry straw yields approximately one kg fresh dhingri. Yields of this mushroom can further be increased by supplementing the substrate with suitable nitrogen source *viz.*, soybean and cotton seed meal.
6. It can be dried and preserved for further use.

(iii) Cultivation of Paddy Straw Mushroom: The Tropical or Chinese Mushroom

Paddy straw mushroom, popularly known as the Tropical Mushroom (*Volvariella* species), appears to be the first amongst the artificially cultivated mushrooms. This is grown mostly in the South-East Asian countries *viz.*, China, Thailand, Philippines, Malaysia, Myanmar etc. Its cultivation was first tried in India sometimes in 1940 and continuous efforts have been made since then to improve upon the cultivation techniques under our own conditions.

Out of various species of paddy straw mushroom, only two are extensively cultivated. These are *Volvariella volvacea* and *V.diplasia*. The fruit body of the *V.volvacea* has a pinkish tinge; while that of *V.diplasia* has whitish hue with a light grey tinge. Both types appear to be very similar to each other. In the initial stages, the fruit body has the shape and size of a bird's egg but soon it swells up and its outer flap ruptures at almost about the middle, leaving a cup shaped structure (volva) at the base. The fruit body is then differentiated into a bulbous base surrounded by a membranous cup (volva), a central stipe (stalk), supporting a cap (pileus), which as it develops assume an umbrella shaped structure. The surface of the fruit body in all stages of development is smooth.

Spawn and Spawn Preparation

Two types of spawn *viz.*, straw spawn and grain spawn are commonly used in its cultivation. Straw spawn is prepared by soaking clean paddy straw cut into 3-5 cm bits. The straw is soaked for 12-24 hours for proper absorption of water. One or two changes of clean water are necessary during the period of soaking. The excess water is drained off from soaked straw by transferring it to a wire gauge sieve. Ten per cent gram dal powder or five per cent oat meal or one per cent wheat/rice bran by weight of the soaked straw, is mixed thoroughly with the straw and the mixture is stuffed in ordinary wide mouthed milk bottles or glucose bottles to two third of their capacity, plugged with non-absorbent cotton and sterilized at 20 lb pressure for 50-60 minutes or at 15 lb pressure for 20 minutes on two consecutive days. After cooling down the bottles are inoculated with a bit of pure culture under aseptic conditions and incubated at 30-35°C. The fungus makes a rapid growth which is visible in the next 2-3 days. The technology for grain spawn preparation of *Volvariella* species is the same as for *A. bisporus* or *Pleurotus* species.

Cultivation Technology

Several techniques have been used for the cultivation of *Volvariella* in the tropics which thrive in the temperature range of 30-35°C and relative humidity 75-85 per cent. However, the traditional method of cultivation in south east Asia is by straw beds, both indoors and outdoor, while the more sophisticated indoor technique is preferred for industrial scale production of the mushroom on cotton waste compost.

Substrate Preparation for Cultivation

Conventional Bed Method

Paddy straw used for cultivation purposes should be fresh, not crumpled or more than one year old and should be free from moulds or green leafy matter. The whole straw is made into bundles, later steeped in running water to absorb moisture, so that it becomes soft at least for 18-24 h. The soaked straw is taken out and kept on a pucca floor in an inclined position to drain out excessive water and later on laid into beds in the following manner:

The soaked straw bundles with butts ends on one side are placed lengthwise pressing each other on rectangular/square bamboo frame supported on bricks. A second layer of straw bundles is placed on the top of the first layer with butt ends on opposite side to make a uniform layer. The loose straw ends extending out of the bed should be cut either with sharp knife or a pair of scissor but should be in no case be turned or tucked in the bed to avoid interference with the emergence of the mushroom. The stack becomes ready for spawning. Before spawning, some quantity of arhar dal or gram dal powder or one per cent wheat/rice bran (on dry weight basis) is sprinkled on the bed at 12 to 15 cm distance from each other (arhar or gram dal powder not to be sprinkled in the case of grain spawn). A small bit of straw spawn is placed on the portions, where the dal powder had been sprinkled all along the margin of straw layer. Third and fourth layers of straw make second layer of bed, but with butt ends at right angles to the earlier spawned layer. The next layers constituting the fifth and sixth would have its butt end as for layers 1 and 2. The last layer is spawned all over and covered with 1.5cm thick layer of straw. Finally the whole bed is as long and broad as the length of straw and about 45-50cm in height. The bed is then pressed tightly so that there are no air spaces which, if allowed, give rise to weed fungi/competitors moulds ultimately interfering with production of the mushroom crop.

Improved Cage Method

In order to prepare beds through cage, select dry, fresh and hand thrashed paddy straw free from moulds and leaf matter. Make 25cm long and 10cm thick bundles @ 60 bundles for each cage. Soak the bundles in boiling water for 20-30 minutes and allow to cool and drain excess water. Disinfect the cage and polythene sheet with 2 per cent formalin solution. Arrange ten straw bundles uniformly in the cage at the bottom layer and put 100gm spawn grain over and inside the bundles. Put up a second layer of ten bundles over the first and spawn in the similar fashion. Repeat till six layers of bundles or till complete filling of the cage. Spray solutions of 0.1′ per cent Malathion and 0.2 per cent Dithane Z-78 all over the bed. Cover with transparent polythene sheet and tie it with sutli. Keep the spawned cage in a shed for mycelial run. A warm lace with temperature above 30°C is helpful for better mycelial growth. Remove the sheet after complete spawn run. Maintain high humidity in the room till pinhead appear. Pinhead appear within 10-15 days after spawning. Harvest at the egg stage. Keep high temperature (32-35°C) and high humidity (over 90 per cent) for the next flush of mushroom to appear with in a week or so.

Industrial Method for Straw Mushroom Cultivation

Chang and his co-workers in Hong Kong have developed a method for straw mushroom cultivation, which is based on a technology similar to that meant for cultivation of white button mushroom, but differs in the ingredients of the substrate as well as temperature and aeration requirements. This also involves fermentation of compost material and pasteurization.

Many agro waste or farm residues such as paddy straw, stems of leguminous plants like soyabean, French bean, cowpea, rice bran and industrial products especially cotton spinning waste, mixed with a little quantity of urea or ammonium Sulphate or CAN and lime are used in the industrial method. Two types of composts are prepared: (1) of slow decomposing material like paddy straw; (2) of fast decomposing material like cotton spinning wastes. These are composted separately.

Slow Decomposing Material

Steep paddy straw in water for 12-24h to soften it. Then mix urea (1 per cent) or ammonium sulphate (1.5 per cent) and heap it with the help of wooden side boards in a rectangular pack. Open the heap after 3-4 days and remake it tightly packed by thorough pressing. This is done to encourage anaerobic fermentation. The height of the heap is generally 1.0–1.5m in both stages. On the 7th and 8th day of the start of composting, this heap is again broken and double superphosphate @ 1-2 per cent of dry weight of straw is added and made into a pyramid about 1 m in height and allowed remain for 2-3 days more.

Fast Decomposing Material

Cotton spinning remnants are soaked in water for 1-2 h. These are mixed with 0.5-1.0 per cent urea and 2-3 per cent chicken manure and made into tightly compact heap 1.0–1.5m height. After 2-3 days, the heap is broken, moisture is adjusted if required and remade into a pyramid one meter high and allowed to remain for 1-2 days. It should be ensured that both types of composts *viz.*, slow decomposing and fast decomposing become ready almost simultaneously. Both the composts are laid in layers to form a heap. The slow growing compost material is mixed with 0.5 per cent lime and 3-4 per cent rice or wheat bran. It forms the basal layer about 12-15cm thick and a thin layer (4-5cm) of fast decomposing compost is placed to form the upper layer. These alternating layers are laid till the heap is one meter high. Layers are placed in a loosened form and never pressed. After about 36 to 48 hour, it is, finally ready for pasteurization.

The final product of a mixture of slow decomposing and fast decomposing material is thoroughly mixed, moisture adjusted and placed on the stacks by light pressing in the pasteurization room. The temperature of the room is raised to 40-45°C by letting the steam in and maintained for 4-5 hour. Ventilation is turned on for 10-15 min. and the room temperature is raised to 63-67°C and maintained for 2 hour. The compost is then allowed to cool by opening the doors and windows till the temperature falls down to 35-37°C. The compost is then spawned, the pinheads will form on the compost. The application of water should be stopped and humidity in the growing rooms is raised to 95 per cent by sprinkling water on the floors and walls, but not directly on the compost beds after 5 days.

The mushroom crop will be ready for picking 2-3 days later and its flushes appear at intervals of 7-10 days (Plate 4)

The total production by this method is reported to be 14-16 kg per square meter bed area. Proper watering and maintaining of congenial temperature within the bed are important factors for the maximum spread of mycelium of the paddy straw fungus. If bed temperature, after 48 hour of spawning remain between 34-38°C, weed fungi, rarely come up.

Harvesting and Marketing

The mushrooms for fresh consumption should be harvested when the volva is about to rupture. Later the stem continues to grow and the pileus (cap) starts expanding but the weight of mushroom almost remains constant. Soon the stem starts loosing its tenderness and is a bit too hard for culinary purposes. The mushroom should be taken out by twisting so that the broken pieces are not left out in the bed. Brine solution (10 per cent) dip is used for preservation of mushroom for a week. Paddy straw mushroom can be easily air dried either in the shade or in sun by spreading on a white sheet of cloth. If the mushrooms are dried in hot air drier, the temperature during drying should not exceed 52°C, otherwise aroma is lost. It is not suitable for canning, because it contains many water soluble constituents responsible for aroma and flavour which are lost in liquid preservation.

(iv) Cultivation of *Calocybe indica*: The White Milky Mushroom

During the last few decade a number of wild edible fungi have been successfully brought under cultivation. *Calocybe indica* is known as milky mushroom or dudh chatta or summer mushroom (Plate-5).

It was reported for the first time in India by Purkayastha and Chandra (1974). It grows in nature on humus soil under the road side trees in forest and sold in city and village markets collected from the forests in West Bengal. It is generally liked by the consumer because of its attractive robust white sporocarps, long shelf life and taste.

Spawn and Spawn Preparation

The spawn of *Calocybe indica* is prepared in the similar manner as is done for *Agaricus bisporus* or *Pleurotus sajor-caju* on wheat grains. After autoclaving, the bottles are inoculated with a mycelial bits taken from 10-15 days old culture. Inoculated bottles are incubated at 26-28°C for 25 days and provided light (6-7 hours/day).

Cultivation Technology

Substrate Preparation for Cultivation

Milky mushroom can be grown on a wide range of substrates as in case of oyster mushroom. It can be grown on substrate containing lignin, cellulose and hemicelluloses. It can be grown on straw of paddy, wheat, ragi, bajra/maize, cotton stalks and leaves, sugarcane bagasse, cotton and jute wastes etc. However, cereal straw (paddy/ wheat) easily available in abundance, is being widely used. The mycelium of *C. indica* can grow either on composted or fresh straw. For preparing compost; soil, sand and maize meal are mixed in 12:6:1 proportion and sprinkled with sterile water. It is then autoclaved at 15 lbs pressure for one hour.

Fresh paddy or wheat straw chopped (5-7 cm) and soaked in clean water for 18-24 hours and subsequently soaked in hot water (80-85°C) for 60-80 minutes is easy and more suitable for growing *C. indica*. Soaked straw is supplemented either with maize meal or wheat bran @ 5 per cent (wet weight of straw). It is then disinfected and filled in wooden trays or polythene bags. Pasteurization of substrate at 65°C for 5-6 hours gives better results.

Spawning and Spawn Running

Spawning rate is kept high and is done at the rate of 4-5 per cent of the wet weight of substrate. Organic supplements like maize meal, wheat bran or rice bran at 5-7.5 per cent concentration may also be added in the substrate at the time of spawning as well as at the time of casing. After spawning, bags are shifted to spawn running room and kept at a temperature of 30-32°C and relative humidity of above 80 per cent. After about 18-20 days when the substrate is fully colonized by mycelium, the bags are ready for casing.

Casing Material and Casing

Casing material (soil 75 per cent + sand 25 per cent) with pH adjusted to 7.5 to 7.9 with 12 per cent calcium carbonate, is pasteurized in autoclave at 15 lbs pressure for one hour or chemically treated with formaldehyde solution (4 per cent) for a week in advance of casing. The fully colonized bags are made uniform and 2-3 cm thick casing material is added to them (about 2kg per bag). These bags are then sprayed with a solution of carbendazim (0.1 per cent) + formaldehyde (0.5 per cent). Malathion at the rate of 0.1 per cent is also sprayed to protect the crop from flies. Regular sprays of the above chemicals should be done on the floor and walls of cropping room.

Cropping and Crop Management

It takes about 10 days for mycelium to reach on the top of casing layer, by which time fresh air is introduced while maintaining proper temperature (30-35°C) and relative humidity (80-85 per cent). Light should be provided for 5-7 hours. Watering is very important to get good and healthy crop. The changes thus made in environment result in the initiation of fruiting bodies with in 3-5 days in the form of needle shape, which mature in about a week.

Harvesting and Marketing

First flush of fruit bodies can be harvested within 3-4 weeks of casing. While harvesting, individual fruit bodies are picked up without disturbing adjoining young pinheads. Mushroom (7-8 cm in diameter) are harvested by twisting. After proper cleaning, are packed in perforated polythene bags for marketing. From one bag of 4 kg wet paddy straw approximately 600-800 g fresh mushroom can be harvested.

Mature fruit body of *C.indica* contains highest protein (17.2 per cent on dry weight basis) while young pinheads contain the lowest protein (15 per cent on dry weight basis). It contains 12 amino acid and glycine is predominant.

(v) Cultivation of *Lentinula edodes*: Shiitake Mushroom

This mushroom is commonly known as 'Shiitake' in Japan and has enjoyed a good international reputation as food. It has taken an important place in the agricultural industry in Japan, Korea and China, but other Asian countries still depend for their supply from forests. At present, its annual production is about 16,29,000 tonnes. It is liked by consumers because of its unique taste and flavour and presence of a chemical which reduces plasma cholesterol level.

Cultivation Technology

Lentinula edodes grows in nature on the dead wood of a number of hard wood trees mainly *Ouercus* spp. (Oak), *Castenopsis* spp., *Elacocarpus* spp., *Lithocaripus* spp., *Betula* spp. and *Carpinus* spp. It is cultivated round the year by adopting suitable strains either those which can fruit at or above 20°C or between 10-15°C and those fruiting around 10°C. The growing technology consists of following steps:

1. Log preparation
2. Spawn preparation
3. Spawning the logs
4. Crop management.

Log Preparation

The mycelium of *L.edodes* is saprophytic and wood rotting fungi. It mainly grows on dried wooden logs absorbing nutrient from the cambium. Although it grows on any size and age of logs, but the logs with 9-18 cm diameter and from 15 to 20 years old tree are most suitable. The most suitable period for felling or cutting the tree is from autumn (December-January) to early spring when the logs contain highest amount of carbohydrates and other organic substances. The logs should contain a moisture content of 45-55 per cent with pH between 4.5 to 5.5. The filled logs are left as such for 25-45 days for bringing the optimum level of moisture that is 40-45 per cent.

Spawn Preparation

Mainly two types of spawn that is saw dust and wood plug spawn is in use. Saw dust spawn is prepared using any of the following formulae:

1. Saw dust (65 per cent) + Wheat bran (15 per cent) + Used tea leaves (20 per cent) + Water (65 per cent).
2. Saw dust (78 per cent) +Sucrose (1 per cent) + Wheat bran (20 per cent) + Calcium carbonate (1 per cent) + Water content (65 per cent).
3. Saw dust (800g) +Rice bran (200g) + Sucrose (30g) + Potassium nitrate (4g) + Calcium carbonate (6g) + Water (2 liters).

Saw dust after proper sieving to remove bigger size wood particles and other impurities is thoroughly mixed with water. When squeezed between fingers, normally one or two drops of water should ooze out. It is then filled in wide mouthed bottles or in polypropylene bags. With the help of a rod, one inoculation hole is made into the centre of the substrate. After plugging with non-absorbent cotton and covering with aluminum foil, it is autoclaved at 20 lbs pressure for 2 hours. The actively growing mycelium of 10 days old culture is inoculated aseptically and incubated for 30 days at 24±2°C.

Wood plug spawn is prepared by inoculating mycelium on small wedge shaped or small cylindrical wood pieces. When the fungal mycelium impregnates the wood pieces, they are ready for inoculation.

Spawning the Logs

The mycelium of *L. edodes* grows at wide range of temperature (5-30°C) but the most optimum temperature is between 20-26°C. To avoid competitors mould, the spawning should be done at low temperature (14-20°C). For inoculation small holes of 1 × 1 cm size and 1.5 to 2 cm deep are made on the logs with drilling machine. The holes are made at a distance of 20-30 cm (long axis) and 6 cm between each row. Saw dust spawn is filled in the holes or wood plug spawn is inserted by cutting out similar size pieces. The saw dust spawn should be kept soft, not tightly pressed and holes sealed with paraffin wax. The spawning should be mostly done in aseptic condition.

Crop Management

Inoculated logs are kept in open at a place where the physical conditions are most favourable for mycelial growth. The inoculated logs are kept in a flat pile, so that there is minimum light exposure and covered with either straw or gunny bags to prevent excessive water loss of the logs. The vegetative growth in the logs will be completed within 8-12 months depending upon the culture strain and the type of wood used.

For fruit body induction it requires temperature shock or temperature drop, high humidity and enough light. For temperature shock, the logs are either sprayed with cold water or immersed in a tank of cold water (in summer for 24 hours in cold water at 15-18°C and during winter for 2-3 days at 10-15°C).

The logs are then leaned against the supports. The cropping area is kept moist to maintain high relative humidity. The temperature should be 15-20°C and humidity around 80-90 per cent. Fruit bodies are harvested by first pressing and then twisting. Mushrooms are harvested upto three times and after a rest for 30-40 days, they are again watered to get more mushroom. It can be repeated upto 3-4 times per year and these logs will produce crop upto 4-6 years (Plate 6).

Synthetic Bag Cultivation

The cultivation of Shiitake by log method requires a very log period (3-5 years). In addition log woods can not be procured easily, being very costly. Therefore, its commercial cultivation can be carried out using saw dust of oak (*Quercus* sp.), maple (*Aser* sp.) or burch. Various steps involved are; substrate preparation, spawning, spawn run and fruiting.

Substrate Preparation

For substrate preparation, various formulation are in use for preparation of substrate which are given as under:

1. Saw dust (80 per cent) + Rice bran (20 per cent) + Water content adjusted to 65 per cent.
2. Saw dust maple and birch 60:40 (80 per cent) + Millet (10 per cent) + Wheat bran (10 per cent).
3. Hard wood saw dust (89.2 per cent) + Rice bran (10 per cent) + Calcium carbonate (0.2 per cent) + Water content adjusted to 60 per cent.
4. Corn cobs (40kg) + Saw dust (10kg) + Wheat bran (12.5kg) + Cane sugar (1kg) + Pectin (15g) + Urea (20g).
5. Paddy straw (50 per cent) + Wheat straw (20 per cent) + Saw dust (20 per cent) + Cane sugar (1.3 per cent) + Calcium carbonate (1.5 per cent) + Citric acid (0.2 per cent) + Calcium sulphate (0.5 per cent).

The water content of the substrate should be adjusted to 60-65 per cent and pH between 5.5 to 6.0. Soluble ingredients (citric acid, sugar, sulphate) are usually dissolved first in water before mixing, saw dust has to be soaked at least for two days and rice straw for three hours. All the ingredients are thoroughly mixed.

The polypropylene bags should be of 15-30 cm size. The substrate mixture is filled in the polypropylene bags and then autoclaved for 1 hour at 121°C. After cooling down, the bags are inoculated with spawn (2-5 per cent) and kept at 20-25°C in dark for 45-60 days to complete the mycelial growth.

When the mycelium in the bag is matured the polypropylene bags are removed. The compressed blocks are then treated for fruiting, either by lowering the temperature (16-18°C) or by soaking in cold water. The primordia appear within 3-4 weeks time and mushroom develops in another 7-10 days. If the temperature in the cropping rooms is in the range of 15-25°C a good quality mushroom crop can be harvested. The relative humidity should be between 85-90 per cent. During primordial formation there is no need of light, but it does play a role in the subsequent fruit body formation. Some of the strains of *L. edodes* with low light intensity (120-200 lux) gave higher yield while with medium light intensity (500-600 lux) they yielded poorly. Low light intensity on the other hand lengthen their cultivation period while medium light intensity stimulates them to fruit more rapidly.

Harvesting, Drying and Storage

Harvesting should be done at an early stage. Normal yields are 15-30 per cent of the wet weight of substrate. Shiitake mushroom can be eaten fresh, dried, canned or picked in vinegar. The dried shiitake should have 10-13 per cent moisture.

(vi) Cultivation of *Flammulina velutipes*: The Winter Mushroom

Although mushrooms are popular all over the world, the cultivation of edible fungi is still restricted to a few species only. Out of which *Agaricus bisporus, Lentinula edodes, Volvariella volvacea* and species of *Pleurotus* are predominant. There is an increasing awareness that waste products especially of agriculture and forestry must be utilized profitably and, therefore, domestication of edible fungi on these wastes would go a long way in alleviating the food crisis and utilizing waste material as well as reducing chances of environmental pollution.

Cultivation Technology

F. velutipes does not require any fermented substrate and can grow well in unfermented material. Different substrates like straw (paddy and wheat), saw dust (Oak wood) and crushed corn cobs (either alone or in mixture with paddy straw) can be used for cultivation of these fungi with success. Wheat and paddy straw should be chopped into small pieces (3-5cm) in order to improve the water holding capacity of straw. The chopped straw is then wetted for 4-6 hour so as to obtain a moisture content of about 65 per cent. The wetted straw should be steam treated (70°C) for 2 hour or can easily be treated with hot water for 3-4 hours to eliminate any chance of contamination. The straw is allowed to cool down to 20-25°C and then spawned. In case of oak saw dust, it has to steamed first at 100°C for 1 hour and then mixed with rice bran in the ratio of 2:1 (w/w) and moisture is adjusted to approximately 65 per cent. Similarly shelled corn cobs are first crushed into small bits and then mixed with cut paddy straw (2:1 w/w) and moisture adjusted to about 65 per cent. Once the substrate has been selected and prepared, it should be spawned at the rate of 50g wheat grain spawn per 500g of dry substrate. The method of spawn production is similar to that used for *Agaricus* and *Pleurotus* species. The cultivation of these edible fungi should preferably be done either in polypropylene bottles (800-1000ml capacity) or wooden trays as these fungi are more prone to contamination. Polythene bags can also be used to advantage. The polythene bags are removed or cut out after the spawn run is complete and then hung in the air. The substrate is filled in the containers and spawned at the rate of 1 per cent of the dry substrate by thoroughly mixing it with substrate. The spawned bags/containers are covered with minutely perforated polythene sheet and optimum temperature of 22-24°C and relative humidity of about 80 per cent is maintained during the spawn run period. Once the mycelium has permeated the substrate, polythene sheet should be removed and the temperature lowered to 10-12°C and relative humidity to about 75-85 per cent.

Environmental Crop Management

The fully colonized bags/containers with spawn, should be carefully sprayed, taking care that the substrate does not start rotting. The maintenance of proper environmental condition (temperature, humidity and aeration) is very essential. The fruit bodies will appear after 12-15 days of keeping the bags under optimal conditions. As soon as small fruiting bodies start appearing on the surface, slight aeration should be provided as high level of carbon dioxide has been found to cause unusually very long stipes and small pilei.

F. velutipes exhibits a definite requirement for light during fruit bodies formation and, therefore light should be provided for 3-4 hours every day. When the fruit bodies are about 2-3 cm long, a support in the form of string or polythene strips should be tied around the stipes so as not to allow the fruit bodies to drop down on elongation (Plate 7).

Harvesting and Marketing

The fruit bodies should be harvested when the stipes are about 8-10 cm long. The harvesting should be done by twisting the fruit bodies in an anti clockwise direction so as to remove minimum amount of substrate, and not leaving any bit of the mushroom on the substrate and at the same time the surrounding fruit bodies should be least disturbed. After harvesting, the lower portion with adhering debris should be cut with the help of a clean knife. They can be consumed either fresh or after drying. Dried fruit bodies can also be used for soup making and flavouring purposes.

(vii) Cultivation of *Agrocybe aegerita*: The Black Poplar Mushroom

The exploitation for domestication of new edible fungi is a most challenging problem because the environmental parameters and physiological factors involved in the control of their morphogenesis and production of fruiting bodies are unknown at the beginning and therefore, require lot of expertise. For example, *Agrocybe aegerita*, commonly known as black poplar mushroom which grows in temperate climates mostly on poplar and willow wood and develop fruit bodies in nature from spring to autumn is a recent introduction. It has uses due to its unique flavour, nutritive and medicinal values that is characterized by low yields.

Cultivation Technology

Preparation of Substrate

It can be grown on straw of paddy, wheat, ragi, maize, bajra, cotton stalks and leaves, sugarcane bagasse, cotton and jute waste. Wheat and paddy straw should be chopped into small pieces (3-5 cm) in order to improve the water holding capacity of the straw. The chopped straw is then wetted for 4-6 hour so as to get moisture content of about 65 per cent. The wetted straw should be steam treated (70°C) for 2 hour or it can be easily treated with hot water for

3-4 hour so as to eliminate the contaminating microflora of the substrate. The straw is allowed to cool down to 20-25°C and then spawned. For using oak saw dust as substrate, it should be steamed at 100°C for one hour and then mixed with rice bran in the ratio of 2:1 (w/w) keeping the moisture content roughly round 65 per cent. Substrate is filled in polypropylene bags (35 × 45 cm, holding 2-3 kg wet substrate) and sterilized at 15 lb pressure for one hour. Once pasteurization/sterilization is over, straw is shifted to spawning room for cooling, bag filling and spawning.

Spawning and Spawn Run

Spawning methods are similar to that mentioned in case of oyster mushroom. In case of sterilization of polypropylene bags in autoclave, either surface or through spawning should be done. Higher spawn dose of 3 per cent of wet weight of substrate is used. Supplements like rice bran or wheat bran at 5 per cent concentration may also be added in the substrate at the time of spawning. After spawning bags are shifted to spawn running room and kept in dark where temperature of 24-26°C and relative humidity of above 85 per cent are maintained. After about 20 days complete spawn run takes place.

Cropping and After Care

After the mycelium has permeated the substrate, polythene sheets are removed and a temperature of 20-25°C and relative humidity of above 80 per cent is maintained. The spawn run substrate should be watered as and when necessary with slight rays, taking care that substrate does not start rotting. The maintenance of proper environmental condition (temperature, humidity and aeration) is very essential. After about 10-15 days fruit bodies start appearing (Plate 8).

As soon as small fruiting bodies start appearing on the surface, slight aeration should be provided as high level of carbon dioxide has been found to cause unusually very long stipes and small pilei. *Agrocybe aegerita* can grow normally even in the absence of light.

Harvesting and Marketing

Mushrooms are easily harvested by hand. In 3-4 flushes, one can harvest 500-600 fresh mushrooms/kg of dry straw. The fruit bodies should be harvested when stipe length is about 8-10 cm, pileus diameter 2cm and weight of fruit bodies is about 2.0–2.5 g. The harvesting should be done by twisting the fruit bodies in an anti clockwise direction so as to remove minimum amount of substrate and not leaving any bit of mushroom on the substrate and at the same time the surrounding fruit bodies should be least disturbed. After harvesting, the lower portion with adhering debris should be cut with the help of a clean knife. The fruit bodies can be stored for more than 10 days in a refrigerator at 4°C. The fruit bodies are easily sun dried with dry matter content of about 10-11 per cent.

(viii) Cultivation of *Auricularia* Species: The Black Ear Mushroom

Among the cultivated mushrooms, Black Ear Mushroom has the oldest record of cultivation by Chinese dating back to 600AD. At present its cultivation has become a major business in some of the South East Asian Countries. Its production has already increased over the years and it contributes roughly 8 per cent of total annual world production of mushrooms (that is 4 million tonnes) and ranks fourth among different mushroom species. All species of *Auricularia* are edible. Among the commercially available species there are three (*A. auricula, A. fuscosuccinia*) and thick (*A. polytricha*) types of mushrooms. The former are light coloured and small, while the latter are dark coloured, large and hairy and do not turn slimy on cooking. This mushroom is believed to cure sore throat, anemia, certain digestive disorders especially piles on regular consumption.

Cultivation Technology

Preparation of Spawn

Two types of spawn *viz.*, grain and saw dust spawn is used. The procedure for preparing grain spawn is similar to that used in case of *Agaricus* or *Pleurotus* species, where as for preparing saw dust spawn, any of the following formulae can be used.

1. Saw dust (65 per cent), wheat bran (15 per cent), used tea leaves (20 per cent) and water (65 per cent).
2. Saw dust (880g), Rice bran (320g), Sucrose (30g), Potassium nitrate (4g) Calcium carbonate and water (2 liters).

All the substrate are thoroughly sieved to remove bigger particles and mixed with water. The substrate properly mixed is, filled in, either in empty spawn bottles or in polypropylene bags and plugged with non-absorbent cotton. These are then autoclaved at 20 lbs pressure for 1.5 hours. After cooling, they are inoculated with 10-20 days old cultures and incubated for 30 days at 25°C.

Cultivation of Logs

Cultivation of *Auricularia* on logs is done in similar manner as that of *Lentinula* cultivation. It can grow on any tree log except that of pines. The logs are cut after autumn with moisture content of 50-65 per cent and with enough sugar accumulation. The tree should be normally 6-10 years old and a log length of one meter with diameter 3-6cm are preferred. For inoculating spawn holes are bored around the logs by means of a chisel or with an electric drill. After spawning the holes are spaced at least 10-15cm apart and are covered with a piece of bark and sealed with melted wax.

The inoculated logs are incubated by laying them in a yard under shade with optimum temperature of 24-26°C. The logs are laid crosswise or stood upright in the yard and covered with plastic sheet to maintain proper moisture. After about 60 days the logs are ready for cropping. They are transferred to the cropping yard which may be an area in the forest or under shade where temperature (22-25°C) and moisture condition (Relative humidity 85-90 per cent) are maintained by frequently spraying logs with water.

Mushrooms can be harvested for about 30 days after exposure to optimum cropping conditions. If the logs are protected properly, harvest can be taken from these logs every year in the natural growing seasons. The total yield after 5-6 years of cropping is reported to be 12-15 per cent of the original weight of the log.

Growing in Bags

Now the cultivation of *Auricularia* is done in polypropylene bags, due to scarcity of wood as well as, ease in handling. Either wheat or paddy straw is used as substrate for cultivation. The cultivation process consists of procuring good quality straw. After cutting them in small pieces of 4-6cm, the straw is soaked in water overnight for 16-18 hours. After draining excess water, the soaked straw is supplemented with 4-5 per cent of wheat or rice bran. The mixture is filled in the polypropylene bags @ 4 kg per bag and autoclaved at 15 lbs pressure for 1-2 hours. On cooling, spawn is added to the substrate, aseptically @ 2 per cent and incubated at 25-

28°C for spawn run for 20-25 days. Slitting of bags for fruiting was found to be more beneficial than complete removal of bags, which prevented drying out of substrate.

Apart from temperature and humidity, light, aeration and watering are other pre-requisite for fruit body development and maturation. Give daily spray, 1 hour diffused light and aeration.

Harvesting

Mushrooms are easily harvested by hand. Mushrooms harvested from wood logs have longer hair, more tough texture, less attractive colour and longer production period than from straw bags (Plate 9).

In three to four flushes, 1000-1400 g fresh mushroom per kg of dry substrate can be harvested. The fruit bodies are easily sun dried with dry matter content of 8-12 per cent. This mushroom has a special quality of retaining its characteristics crispness on cooking.

Maintenance of proper hygienic conditions are necessary because of addition of supplement like wheat/rice bran, may attract moulds. Presence of high level carbon dioxide results in development of abnormal fruit bodies with long stalks and small pileus. The absence of light also produces abnormal under developed mass of fruit bodies.

Chapter 8

Pests and Pathogens of Mushrooms

Like crops, mushrooms are also subjected to a number of biotic and abiotic factors, which are parasitic to mushroom or its mycelium. Among the biotic agents are fungi, bacteria, viruses, nematodes, insect-pests, mites and springtails which cause damage to mushrooms directly or indirectly, whereas abiotic factors consists mainly of fluctuation of temperature and relative humidity, low and high moisture in the substrate, pH, CO_2 concentration in the room, wind velocity and fumes which make the substrate non-selective for mushroom mycelium. Since the scenario of mushroom cultivation in India is different from all the other mushroom growing countries of the world, with respect to substrate preparation, casing mixture and system of cultivation, thereby making the crop more prone to attack of pests and pathogens. Use of unpasteurized compost, coupled with unhygienic conditions helps in the perpetuation of various pests and pathogens besides resulting in the reduced level of production. The detailed description of various pathogens and pests encountered during the cultivation of various mushrooms and their management is as under:

Fungal Diseases of White Button Mushroom

Cobweb: Mildew or Softdecay

Casual Organisms

Cladobotryum dendroides (*Syn Dactylium dendroides*) imperfect state of *Hypomyces rosellus*.

Symptoms

The mycelium of the pathogen is greyish to white, when young, but reddish as it ages. The fungus grows rapidly over the surface of the casing soil and envelops the fruiting bodies in a cropping beds. The mushrooms appear like cotton balls on the surface but underneath, the mushroom are reduced to a soft rotting mass.

Epidemiology

Soil and air are the common sources of infection and wet surface and high humidity are the predisposing factors. The fungus spreads through condia due to negligence during picking, because left over pieces of mushroom can initiate infection.

Control

1. Thorough disinfection of casing soil, either by live steam or by sterilization at 65°C for 4 hours completely eradicate the pathogens.
2. Reduce relative humidity to 80 per cent and run fan soon after spraying as a preventive measure.
3. Spraying the beds with 50ppm Carbendazim or Benomyl control the disease without inhibiting mushroom mycelium. It can also be best controlled by application of Dithane Z-78 (0.25 per cent) sprayed three times at an interval of 10 days.
4. Treatment with 20 per cent Terraclor (PCNB) dust or dusting localized areas with 70 per cent calcium hypochlorite has also been recommended.

Dry Bubble: Brown spot, fungus spot, dry bubble etc.

Casual Organism

Verticillium fungicola

Symptoms

The most characteristics symptoms is the light brown superficial spotting on the cap. These spots extend in diameter by coalescing into brown patches and are distinctly different to those produced by bacterial blotch. Spots in cap start sinking and centre is covered with a heavy spore mass of dark brown colour with age. In case of severe attack, downward splitting of the stem giving a shattered appearance is also witnessed. Interference with proper growth of stem causes the cap to remain small tilted and misshapened.

Epidemiology

Primary source of infection appears to be the spent compost lying with in the premises of the cropping room and disease spread through, phorid flies mites, and pickers hand. Temperature above 18°C and relative humidity above 90 per cent are predisposing factors.

Control

1. Use of sterilized casing soil, proper disposal of spent compost, control of insect-pest and better hygiene are essential to avoid primary infection.
2. Three sprays at 10 days interval with 0.25 per cent Dithane Z-78/Sporgon, 1.5g/m^2, 9 days after casing reduce infection. Spraying with 0.05 per cent Benlate, also helps in control of the disease.
3. Sanitation practices which include disposal of diseased sporophores after dipping in 5 per cent solution of formalin, disinfection of tools and other implements, sterilization of used trays at 70°C for 10-12 hours and control of mites and flies in the cropping room help in the prevention of the disease significantly. In case of disease prevalence, the crop should be harvested first from healthy cropping rooms and end up with diseased ones.

Wet Bubble: White mould, Wet bubble, La mole, *Mycogone* disease etc.

Casual Organism

Mycogone perniciosa

Symptoms

When this fungus attack mushroom a white, felt-like mould grows on them and rapidly reduces them to soft, white putrescence exuding brown bubble. A swollen stalk and disproportionately small cap indicate infection at an early stage in the development of mushroom (Pate 10).

Epidemiology

Mycogone is a soil inhabiting fungus and the casing soil is the primary source of infection, but introduction of spores by flies should not be over looked.

Control

1. Good hygiene, use of properly sterilized casing soil and spraying with Dithane Z-78 (0.25 per cent) at 10 days interval gives good control.
2. Spraying of Benomyl at 0.5–4g/sq. meter immediately after casing has been reported very effective for the management of wet bubble.
3. Application of Prochloraz manganese complex (Sportak 50WP) into casing mixture have been reported very effective for the management of wet bubble.
4. For general spraying chlorinated water should be used.

Green Mould: *Trichoderma* spot, *Trichoderma* blotch, *Trichoderma* mildew

Casual Organisms

Trichoderma viride, Trichoderma hamatum, Trichoderma koningii.

Symptoms

Green mould develops in seeded bed due to the defective compost and development of toxic substances in the formalin treated compost. The fungal mycelium starts disappearing in patches where the mould grows and a dark grey brown patch with white margin results from this reaction. Some species also attack the fruit bodies producing dark brown or greenish lesion/spots/discoloration of the stipe (Plate 11).

Epidemiology

Wet surface, high humidity, low compost pH (below 7.0) and presence of dead mushroom tissue are primary source of infection. Further spread of spore is by air movement, contaminated equipments, flies and mites.

Control

1. Proper pasteurization of compost and casing mixture. Maintain 7.0 to 7.5 pH and 69-70 per cent moisture in the compost beds.
2. Spray infected beds with 0.05 per cent Dithane Z-78 or 15 per cent Calcium hypochlorite solution.
3. Maintain a strict hygienic condition in and around mushroom house and use correct concentration of formalin (2 per cent) and avoid excessive use of formalin.

Competitors Moulds

The mycelium of the edible mushrooms has also to struggle against competitors which either reduce the available food or retard the growth.

Brown Plaster Mould

Casual Organism

Papulaspora byssina

Symptoms

The mycelium in initially white, which changes later through shades of tan light brown to cinnamon brown colour. It extends outwards from the centre on exposed compost surface and also on casing soil. Mushroom appearing from such beds give delayed crop and yield is reduced even up to 20 per cent.

Epidemiology

Infection starts initially from the infected compost or air. Wet compost, in inadequate ventilation and high temperature during spawn run (28-30°C) and cropping (above 18°C) also favour the spread of this mould.

Control

1. Maintain strict hygienic conditions.
2. Localized treatment of infected spots with 4 per cent formalin is recommended.
3. Excessive moisture should be avoided in the compost.
4. Spraying of Bavistin at the concentration of 0.05 to 0.1 per cent has also been recommended.

False Truffle

Casual Organism

Diehliomyces microsporus

Symptoms

Small tufts of white to cream coloured mycelium appear on the compost or beneath the casing soil. Subsequently, white, wrinkled, round to irregular masses, resembling to brain like structure are formed. Ascocarps of the fungus which are initially spherical to irregular white to creamish in colour, turns into reddish brown colour when mature. A characteristics sulphur or chlorine like smell is present in infected rooms. The fungus does not allow the mushroom mycelium to grow and compost turns to dull brown in colour. The spawn in affected patches turns soggy and disappears (Plate 12)

Epidemiology

The major source of infection are casing soil and surviving ascospores/mycelium in spent compost or in wooden trays from previous crop. Ascospores can survive for a period of five years in soil and spent compost and thus serves as a major source of primary inoculum.

Control

1. The compost should be prepared on cemented floor. The pasteurization and conditioning of compost should be carried out carefully.
2. Temperature above 26-27°C during spawn run and after casing should be avoided. During cropping temperature should be kept between 14-15°C.
3. Compost should be cooked out at 70°C for 12 hours at the end of the crop which will completely kill the mycelium

and spores of pathogen in the compost. Wooden trays should be separately chemically sterilized.

4. Initial infection can be checked by treating the affected patches with a solution of 2 per cent formalin.

Olive Green Mould

Casual Organism

Chaetomium olivaceum

Symptoms

Greyish white mycelium with a small round, olive green pinhead like bodies (perithecia) are attached to the finished straw pieces of the compost. Spores of the mould are heat resistant and quite abundant in population in the compost and casing soil or even in spent compost. Spores germinate on wet compost and fungus spreads rapidly under low levels of oxygen.

Epidemiology

The infection usually comes through air, compost and casing soil. Compost pasteurized at a temperature exceeding 60°C in the absence of adequate fresh air pre-dispose the compost to infection.

Control

1. Introduce ample fresh air without allowing bed temperature to exceed 60°C during pasteurization.
2. Spray the beds with 0.2 per cent solution of Thiram or 0.5 per cent solution of Benlate as check against further spreads. Sprays of Dithane Z-78 (0.2 per cent) have also been recommended for checking the secondary spread.
3. Re-spawn the beds in case of high incidence of the mould.

Yellow Mould: Mat disease, Vert-de-gris

Casual Organism

Myceliophthora lutea, Sepedonium chrysospermum, Chrysosporium sulfureum

Symptoms

Small colonies of yellow, browny yellow or yellow green moulds are often found in the compost, towards the end of the crop.

Myceliophthera lutea has been reported to induce 'Mat disease' in which a corky mat forms a seal between compost and casing.

Sepedonium chrysospermum and *Chrysosporium sulfureum* are mainly observed in the compost and are initially white in colour, turning to yellow or tan at maturity. The toxins produced by the mould inhibit the spawn run and ultimately the mushroom mycelium disappear from the compost (Plate 13).

Epidemiology

Primary source of infection is spent compost, casing soil and chicken manure. Secondary spread is mainly through mites, flies water splashes etc.

Control

1. Proper pasteurization of compost and casing.
2. Chicken manure should be properly sterilized separately either with 2 per cent formalin or 0.25 to 0.5 per cent Carbendazim.
3. Benomyl (0.05 per cent) and Blitox (0.04 per cent) sprays have been found quite effective to control the disease.
4. Strict hygienic condition should be followed.

Ink Cap: Weed mushroom, Ink weed

Casual Organism

Coprinus atramentarius

Symptoms

Long slender white stalks with small thin caps often appear on the compost beds before casing. The development is rapid and caps decay into a black liquid.

Epidemiology

Under composting and presence of ammonia in the compost are considered major factors for their appearance.

Control

1. Proper pasteurization of compost and casing and elimination of ammonia fumes at the time of filling of the trays should be ensured.

2. Avoid excessive watering.
3. Mechanical pulling out of *Coprinus* from the beds is quite effective. However in case, the number of fruit bodies (*Coprinus* spp.) exceed 200 in the initial stages, the compost should be re-pasteurized, re-spawned and re-cased.

Cinnamon Mould

Casual Organism

Chromelasporium fulva

Symptoms

The mould first appear as large circular patches of white aerial mycelium on the compost or casing. As the spore mature the colour changes to golden brown or cinnamon and the colony develops a granular appearance.

Epidemiology

Casing soil mixture is the primary source of infection. Over pasteurized compost, over heated patches during spawn run, high moisture content of the compost and excess of ammonia present in the compost favour the disease development.

Control

1. Proper pasteurization of casing soil.
2. Spray newly cased beds with Dithane Z-78 @ 0.25 per cent.
3. Maintain proper moisture content in casing layer.

Lipstic Mould

Casual Organism

Sporendonema purpurescens

Symptoms

The disease first appear in spawned compost as a white crystalline like mould which is very difficult to distinguish from spawn. As the spore mature, the colour changes from white to pink to cherry red and then to dull orange or buff.

Epidemiology

Casing soil and spent compost are source of primary inoculum. Further spread is through water splashes and by pickers.

Control

1. Maintain strict hygienic condition.
2. Proper pasteurization and conditioning of compost will eliminate the fungus.

White Plater Mould

Casual Organism

Scopulariopsis fimicola

Symptoms

The disease appear as white patches on the compost or casing soil. These patches or mycelial mats may be more than 50cm under favourable conditions. The white growth changes to light pink after a week of the formation of the spot. Spawn run is reduced significantly and under severe conditions complete failure are also recorded.

Epidemiology

The mould is favoured by high pH of the compost (above 8.0) and in compost which still retains ammonia smell.

Control

1. Proper composting and addition of optimum quantities of water and gypsum are recommended.
2. Spray of Benomyl (0.1 per cent) and localized application of formalin (2 per cent) after the removal of the mat are helpful in controlling the mould.

Lilliputia Mould

Casual Organism

Gliocladium prolificum

Symptoms

Mould seriously restrict the spawn spread in the compost resulting in poor yield.

Epidemiology

Chicken manure, horse manure and casing mixture are the primary sources of infection.

Control

Use of Dithane Z-78 at 20 ppm concentration has been recommended for the control of the mould.

Pink Mould

Casual Organism

Cephalothecium roseum

Symptoms

Mould appears as white growth on the casing soil which turns pink in due course.

Epidemiology

Primary source of infection is through air.

Control

Mould can be checked by spraying twice with Thiram or Captan (0.05 per cent) on casing soil at 10 days interval.

Fungal Diseases and Competitors Mould of *Pleurotus* Species

Fungal Diseases

In all nine fungal pathogens have been reported attacking fruit bodies of different *Pleurotus* species. Most important of these fungal pathogens are *Verticillium fungicola, Gliocladium deliquescens, Sibirina fungicola* and *Cladobotryum verticillatum.*

Sibirina Rot

Symptoms

White cottony growth on the substrate, small green to brown irregular spots or fluffy growth on fruit bodies, soft rot and decay of sporophores emitting foul smell. *Sibirina fungicola* induces powdery white growth on stipe, gills and primordial.

Control

1. Regular spray of Bavistin (50ppm to 100ppm) at an interval of 15 days.

2. Providing proper aeration and maintaining proper relative humidity.
3. Strict hygienic condition in the sheds.

Competitors Moulds

Compared to white button mushroom the information on diseases and competitors mould occurring in or on oyster mushroom are less. Several competitors moulds have been reported occurring in the substrate used for oyster mushroom cultivation. Variations in the number and types of mould are mainly due to the use of a variety of substrate, different methods of substrate preparation and the conditions and containers used for cultivation. Most of the competitors mould have been reported to be completely inhibited under *in vitro*/or *in vivo* conditions by spraying Benomyl (50ppm), Carbendazim + Blitox (100ppm each) and Thiram (100ppm).

Fungal Diseases and Competitors Mould of Other Mushroom Including Paddy Straw Mushroom (*Volvariella* Species)

Paddy straw mushrooms are subject to a number of destructive diseases/competitors mould like *Mycogone perniciosa, Verticillium* spp. etc. A "button rot" disease caused by *Sclerotium* species have also been reported. Combination of insecticide, fungicide and antibiotic (Malathion 0.025 per cent + Dithane Z-78 or Binomyl 0.025 per cent + Tetracycline 0.025 per cent) are recommended for the management of pests and diseases of these mushroom. Several other competitor moulds namely; *Coprinus aratus, C.lacopus, Penicillium* spp., *Aspergillus* spp., *Rhizopus* spp., and *Selerotium* spp. have been reported from the substrate. Partial sterilization of the straw and sprays of Captan and Zineb (0.2 per cent) on beds have been recommended for reducing the damage *Cladobotryum verticillatum* has been recorded on *Auricularia polytricha* producing white fluffy growth on substrate and fruit bodies. Spraying Carbendazim (50 ppm) has been reported effective for controlling the disease. Similarly several competitors mould have been reported during the cultivation of *Calocybe indica*, namely *Cladobotryum, Aspergillus niger, A. fumigatus, Trichoderma viride, Fusarium* and *Coprinus* species. Proper sterilization of substrate can reduce the incidence of these moulds.

Bacterial Diseases of *Agaricus* Species

Bacterial Blotch

Casual Organism

Pseudomonas tolaasii

Symptoms

Circular yellowish spots develop on the cap or near the margin and coalesce to form chocolate brown spots which penetrate into the fleshy tissue. In severe cases brown lesions develop even on the stem (Plate 14).

Epidemiology

Inadequately sterilized soil and contaminated implements are the main sources of infection. The pathogen spread through splashing of water drops from infected to healthy sporophores. Pickers, implements flies and mites also help in the spread of the disease to healthy trays.

High humidity (above 85 per cent) and inadequate ventilation during cropping permit the pileus to remain wet for longer periods, which helps in the disease initiation and spread.

Control

1. Lowering of humidity to 80 per cent and running fans immediately after watering to dry the caps prevents bacteria to spread on the growing sporophores. Spray the beds with 100 per cent bleaching powder.
2. Application of terramycin 9mg per square foot and streptomycin (200ppm) has been found effective in managing the disease.

Ginger Bacterial Blotch

Casual Organism

Pseudomonas gingeri

Symptoms

The disease is characterized by the ginger colour of blotches which do not change with age and distinguish this disease from brown blotch. The blotches can never be deeper than 1-2 mm.

Epidemiology

Similar to brown blotch and casing material is probably the most important primary infection.

Control

Similar to brown blotch.

Drippy Gill

Casual Organism

Pseudomonas agaricii

Symptoms

The disease is more severe in autumn and winter months. Gills are attacked by the bacterium before the veil of the mushroom is broken. Affected gills are often under developed, showing small brown decaying areas with creamy white bacterial ooze on them, hence the name drippy gill.

Epidemiology

Probably flies, pickers and water splash spread the disease within the crop.

Control

Similar to those recommended for bacterial blotch.

Pit Disease

The cause of the disease is not yet established but it is suspected to be bacterium. Mites and nematodes have also been associated. Small dark often black and slimy pits appear on the cap surface. The depth of the pits may be few to several mm deep and may vary in number. They are most common in 3rd and 4th flush. Same control measure, as recommended for bacterial blotch may reduce the incidence and spread of the disease.

Bacterial Diseases of Oyster Mushroom

Yellow Blotch

Casual Organism

Psudomonas agarici

Symptoms

The disease appear as blotches of varying sizes on pilei, sometimes depressed, yellow hazel brown or orange in colour. Infected fruit bodies turn yellow and remain stunted under high humidity (more than 90 per cent) the infected fruit bodies show sliminess (Plate 15).

Epidemiology

The pathogen can easily spread inside the mushroom farm through water splashes, workers, tools and mushroom flies. Presence of water film on the surface of fruit bodies is quite favourable for earlier appearance of symptoms.

Control

1. Proper ventilation and careful watering coupled with monitoring of temperature in mushroom unit keep in limiting the disease incidence.
2. Regular application of chlorinated water containing 100-150ppm of freely available chlorine at 3 to 5 days interval help in minimizing the incidence of disease.
3. Use of Oxytetracycline and Streptocycline have also been reported.

In addition to this bacterial rot of *Volvariella* spp., browning of *Lentinula edodes* and brown soft rot of *Flammulina velutipes* have also been reported. The control measures are the same as recommended for yellow blotch of *Pleurotus* spp.

Viral Diseaes of Mushrooms

Mushroom crop is also subjected to attack by a number of viruses which cause disease commonly known as La France, Watery Stipe, Die back, X-disease or brown disease, which may result into slight or complete failure of the crop. On the basis of the size and shape of patches, five different viruses have been reported attacking mushrooms and are known as Virus 1, 2, 3, 4, and 5. These viruses may occur alone or in any combination. In India, virus and virus like disease have been reported on button and oyster mushroom. A double standard RNA polyhedral virus has also been reported in *Volvariella volvacea* from China. Virions of different shapes and sizes have been detected in *Lentinula edodes* from Japan and USA.

Symptoms

The most common symptoms are the elongation of the stalk with a small, titled cap (drumstick). Deterioration of the mycelium (dia-back) is common, which increases with time, resulting into baron patches of the crop. Sometimes small brown mushroom develop which often open prematurely. Affected fruit bodies have a water soaked appearance (water stipe) which are found to be totally water logged when squeezed.

Epidemiology

The viruses cannot live in soil or outside the host tissues as these are obligate parasites. Hence affected crop debris, mushroom mycelium and spores survive mostly on wooden boxes and in the spent compost, which can be easily carried by air to distant places. Transmission of viruses through mushroom spores have been demonstrated. Some strains of mushroom spawn are also known to contain virus particles.

Control

1. Strict hygiene inside the farm.
2. Use of filtered air inside the peak heating, spawn running and cropping rooms.
3. Mushrooms should be picked before they open.
4. All wooden parts of growing units should be thoroughly cleaned and sterilized to kill any mushroom mycelium from the earlier crop with 2 per cent sodium pentachlorophenate.
5. Use only tolerant or resistant strains.
6. Immediately after spawning use a pesticide against flies and cover the compost with paper. Wet the paper twice a week with 2 per cent solution of the 40 per cent commercial formaldehyde.
7. At the end of the crop, cook out the compost inside the room at 70°C for 12 hours with live steam.

Pests of Mushroom

Mushroom flies, mites, springtails and eelworms (nematodes) are some of the major pests invading the mushroom beds.

Mushroom Flies

Three types of flies have been found to cause damage to mushrooms from time to time *viz.*, sciarid, phorids and cacids. These are small, delicate, black, yellowish or brown in colour with different types of wing venations and size. Larval stage is the most damaging. Adults do not cause much damage except help in breeding to increase the population and dissemination of disease spores etc. A brief account of these flies is given hereunder:

Sciarids Flies

These flies have dark slender bodies and long antennae. They themselves cause less damage, but act as carriers of disease spores and of smaller pests, but in addition, they lay eggs in the beds, ultimately producing larvae. These larvae feed on manure making it soggy, devour the mycelium in the compost causing mushroom above to die; and burrow into the base of the stalks and right through to the caps. They also enter the fruiting bodies and make tunnels and honey comb the pileus. Most damaging species are; *Lycoriella fenestralis, Bradysia coprophila, L. mali, L. agarici and Sciara pulla* (Figure 6).

Phorids

Phorids or manure flies are stronger than Sciarids but two-thirds of their length. They have short antennae and are often more

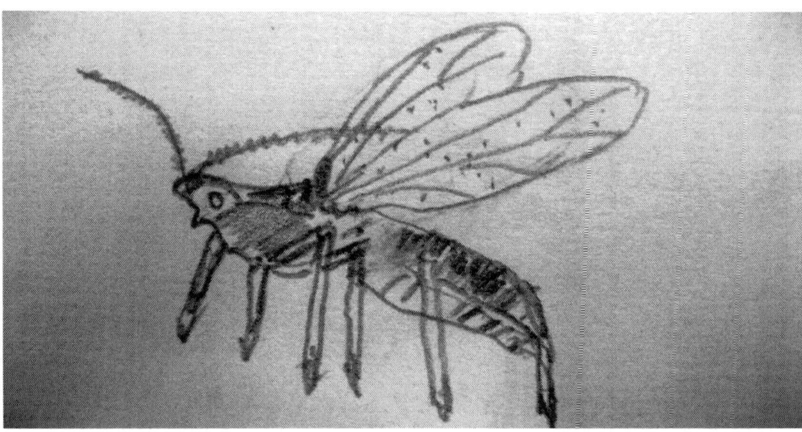

Figure 6: Sciarids Fly

prevalent in summer months. The damage to mushroom beds is done by their larvae, which feed on the spawn or mycelium, pierce the hyphal and suck the contents. Species include, *Megaselia nigra, M. agarici, M. bovista* and *M. halterata* (Figure 7).

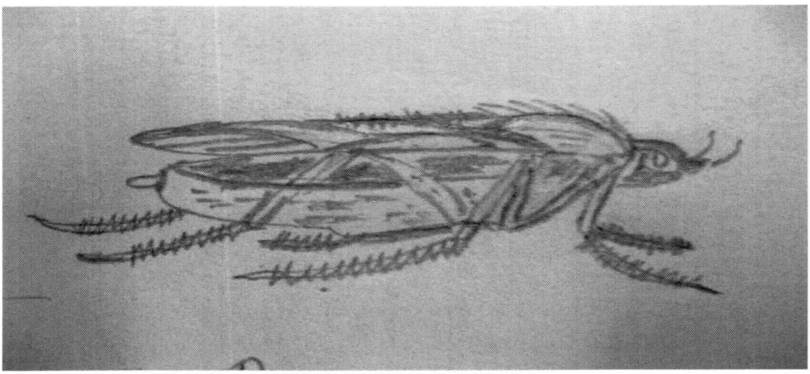

Figure 7: Phorid Fly

Cecids

Cecids flies or gall gnats have long been considered as major pests of the mushroom crop. The tiny flies, which are orange and black in colour are rarely seen. Cecids are smaller than phorids and sciarids, measuring just above 2mm in length. They are of various colour *viz.*, pink-yellow, orange or white pointed at both ends. Cecids larvae not only feed on mycelium but also enter the stalks and into the gills. The most common cecid fly is *Heteropeza pygmea* with white larvae having a spear-shaped anchor process in the pupa larva stage. *Mycophila speyeri* and *M. barnesi* are often seen and their orange larvae are more common than is generally conceded (Figure 8).

Epidemiology

Spent compost, soil and dirt are the breeding centres for these flies, where from these migrate to the mushroom beds either directly or indirectly with the compost, casing and other material. The female flies are attracted by the smell of the nitrogenous material added to the compost and lay eggs and thus contaminate the beds. Under unfavourable conditions, the larvae remain hidden in the cracks and crevices of the wooden trays.

Figure 8: Cecid Fly (*Mycophila speyeri*)

Control

1. Strict hygienic condition in the mushroom house.
2. Proper pasteurization/sterilization of compost and casing soil, mixing of Diazinon/Lindane in the compost (long method of composting) and spraying Diazinon (Ditaf) @ 0.003 per cent at the time of filling are good preventive measure.
3. Malathion (Cythion) or Nuvon (DDVP) may be sprayed alternatively (0.05 per cent) at 3-4 days intervals on the beds and rooms during spawn running period and also at the time of casing. During cropping either, malathion or Nuvon (0.05 per cent) should be sprayed at 7-8 days intervals and mushrooms should not be harvested for 2-3 days in order to reduce the effect of the insecticide.
4. Application of Chlorfenvinphos, Fenitrothion, Fenthion at 1.0g a.i/m^2 to bed surface immediately after spawning and before casing gives satisfactory control.

Mites

Several species of mites have been found associated with the mushroom crop. *Tyrophagus putresentiae* (Figure 9) has been recorded as the most damaging pest under our conditions. *T. putresentiae* feeds on mycelium and resulting in small irregular pits on stalks and

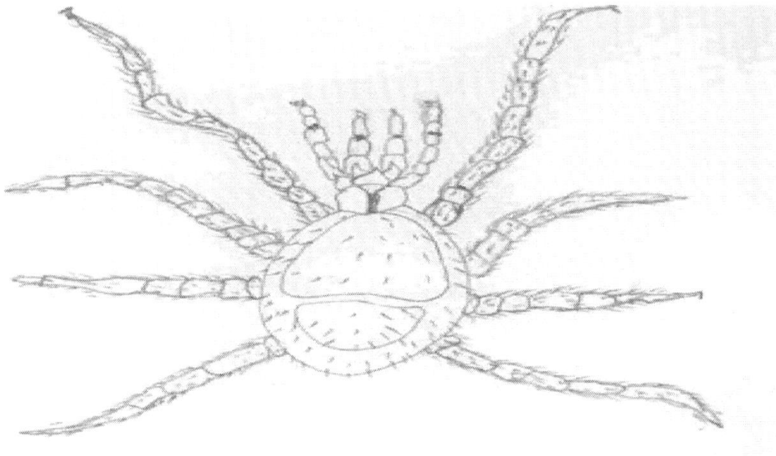

Figure 9 Mite: (*Tyrophagus* spp.)

caps of sporophores. Mites are small, about the size of a pinhead, with white yellowish brown or red colouration and are often seen running speedily over the surface of the mushroom beds, fruit bodies sides of trays, walls and floors of the mushroom houses. In some cases these have glistening bodies and also long hairs.

The harmful mites damage the crop directly by feeding on the spawn and mycelium or puncture holes in mushroom caps and stalks and also cause stunting of fruit bodies as well as brown spots on the caps and stems. Other mites do not damage the mycelium directly as almost all types feed on weed moulds and sometimes create difficulty for workers as they crawl over their body. They also carry spores of several diseases from bed to bed.

Epidemiology

Improperly pasteurized compost and casing soil, old infected beds, spent compost, contaminated implements and mushroom flies are the main sources for carrying mites to mushroom houses.

Control

1. Pasteurize compost and casing material properly.
2. Maintain strict hygienic condition in and around mushroom house.

3. Spray Diazinon (Ditaf) @ 1.5–2.0ml/10 liters of water, outside the mushroom house premises, wash empty trays, wall and floor of the cropping room and composting platform with 0.02 per cent Kelthane solution.
4. Spray 0.01 per cent solution of Kelthane on the mushroom beds during cropping in case of mite infestation.
5. Cook out the exhausted compost in the trays with steam at 70°C for 10-12 hour.

Springtails

Several species of springtails have been reported to damage button, oyster and shiitake mushroom, particularly *Lepidocyrtus cyaneus* (Figure 10). These are small insects (1mm long) with stout antennae and are silvery, reddish brown or dark grayish in appearance. They are active in the dark and remain hidden either under the casing soil compost or fruit bodies and move by springing several centimeters, when disturbed. These tiny insects feed mostly on the mycelium, but also cause damage to caps and stems of fruit bodies.

Epidemiology

Use of inadequately sterilized casing soil and dirty water during composting are the two main sources of springtail infestation.

Control

1. Maintenance of hygienic condition in and around mushroom house.

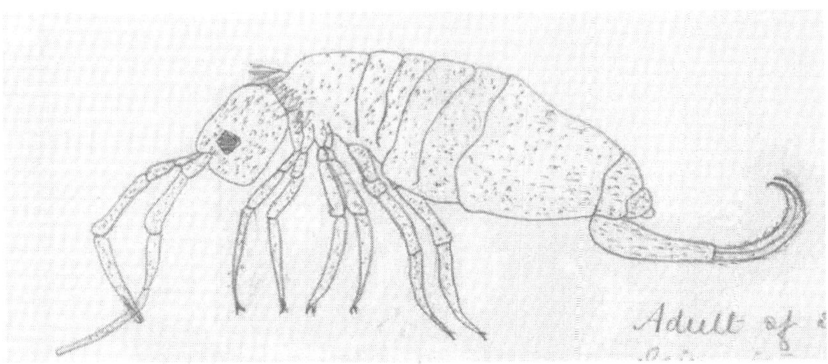

Figure 10: Mushroom Springtails

2. Proper pasteurization of compost and casing material.
3. Spraying with Malathion or Nuvon @ 0.05 per cent on effected beds during cropping is very effective.

Nematodes

Nematodes have often been observed to infest the mushroom beds and reduce productivity of compost and its market value. Nematodes infestation in mushroom are unique in a way that the mushroom nematodes have not only adapted themselves fully with ecological requirements of the crop and multiply very fast to occur in high number in mushroom beds, but these are only examples of ectoparasitic forms, capable of inflicting 100 per cent crop losses. These are small microscopic threads like micro-organisms which can be found in almost, in any type of environment, all types of soils and areas, where organic matter is found and are capable of rapid multiplication (Figure 11)

Nematodes encountered in mushroom compost/beds have been categorized as myceliophagous, saprophagous predaceous, plant parasites and animal parasites on the basis of their nature of parasitism. Among these only myceliophagous and saprophagous forms hold significance as they effect the crop yield through direct feeding or indirectly. Predaceous forms usually occur in low counts and are beneficial as they feed upon the harmful myceliophagous

Figure 11: Mushroom Nematodes

and saprophagous nematodes. Plant parasitic and animal parasitic nematodes occur merely as contaminants.

Myceliophagous nematodes (which feed on mycelium) are known to be the most damaging. These have needle like structures known as stylet on their mouth part, with which mycelial wall is punctured. These nematodes suck the sap and leave it devitalized. Common examples are *Ditylenchus myceliophagus*, *Aphelenchoides agarici*, *A. composticola*, *A. myceliophagus* etc.

The saprophytie or saprophagous nematodes like *Rhabditis*, *Panagrolaimus*, etc. also cause direct injury to the mycelium. These feed on organic matter and disseminate bacterial pathogens besides producing toxins.

Symptoms

The myceliophagous nematodes affected beds shows various types of symptoms like mycelium disappearance, browning and stunting of water soaked pinheads and the fruit bodies appear in patches. The compost emits a particular odour and yield decline drastically.

Epidemiology

Old infested beds, spent compost, in adequately sterilized casing soil, infected compost and casing soil, spray water from natural resources are some of the principal sources for nematodes infestation. These are further disseminated through workers, faulty spraying of water, uncleaned implements and of course the mushroom flies. Nematodes enter into "anabiotic" form under dessicated conditions and become inactive as immobile as dead particles and thus escape unfavourable conditions like high heat, cold and dry conditions and later regain activity in humid conditions.

Control

1. Maintain strict hygienic conditions in the growing rooms by providing doors fitted with fly proof wire mesh and complete sanitation around the mushroom farm.
2. The peat heat temperature during pasteurization must be maintained at 58-59°C for 4 hours.
3. The casing soil should be sterilized properly either with steam heat (65°C) for 5-6 hours or with formalin.

4. The spraying water should preferably be chlorinated and free from nematodes.
5. Wooden trays should be sterilized properly either with steam, formalin or dipped in hot water (80°C) for 2-3 minutes. Alternatively, crops can be grown in polythene bags instead of wooden trays and properly washed in 2 per cent solution of formalin.
6. The exhausted compost should be cooked out with steam at 70°C for 10-12 hour, so as to avoid carry over of the information to subsequent crops.
7. The nematodes trapping fungi like *Arthrobotrys* spp. are also being used as biological control agents for controlling mushroom nematodes. The fungus traps the nearby nematodes and kill them.
8. Some strains of *A.bitorquis* like K-32 have been found to be moderately resistant whereas *P.sajor-caju* was found to the attack of mushroom nematodes. Hence growing of these crops in rotation with the crop of common white button mushroom has been recommended so that inoculum level of the nematodes in the mushroom house is very much reduced.

Abnormalities/Abiotic Disorders of Mushrooms

Stroma Formation

Sometimes white cottony strands are formed into a thick compact mass on the casing soil and hinder the formation of pinheads and fruiting bodies. These conditions can be avoided by lowering the bed temperature and supplying the fresh air as soon as the cottony growth appears. Excessive CO_2 and prolonged spawn run period also results in stroma formation, which ultimately reduce yield. The early patch appearance of the mycelial growth on the casing layer should be covered with freshly sterilized casing soil, so as to induce fruiting.

Rose Comb

Malformation of caps and gills resembling a rooster's comb is often seen in mushroom beds. These misshapen gills make the swellings look spongy. Smoke and oil from coal, kerosene oil or some coal tar products cause the above abnormality, when fumes come in contact with the fruit bodies.

Long Stemmed Mushrooms

Lack of sufficient ventilation or carbon dioxide accumulated in the cropping rooms results into abnormally long stemmed mushroom with small caps. With the improvement of aeration such conditions can be avoided.

Cracked Mushrooms

High temperature followed by less humidity or dry atmosphere inside the cropping rooms cause cracking of the mushroom surface. Humidity in the cropping beds should preferably be maintained between 85-95 per cent to overcome this problem. Asbestos sheet should be fixed in the inner side of the room, in order to avoid direct heat penetrating the beds that cause moisture loss and cracking. Scale arise through the surface tissue, failing to grow, while the caps develop further (Plate 16)

Strong vapours of formaldehyde or pest control products in excess can also cause the outer layer of the skin of half grown mushroom to tear off.

Hard Gill or Open Gill

This is the most common form of malformation because the gills are rudimentary and sometimes merely pale corrugation. It is also referred to as "Open Veil' for the veil too is fragmentary or non-existent. The mechanisms to be the hard gill is genetic and certain strains have a greater tendency to develop the abnormality. Frequent fluctuation in temperature and humidity inside the cropping rooms result in formation of hard gilled mushrooms in which veils and gills are not formed.

Early Opening of Mushrooms

High bed temperature (above 22°C) and dry compost below the casing layer are the pre disposing factors for mushrooms which open up at premature stage.

Leakers/Weepers/Stinckers

Mushroom described as being 'Weepers' typically exude considerable amount of water from mushroom cap. When small water droplets exude from the stem or cap, the mushrooms are called leakers. Water collects on the casing surface beneath a weeper and the area develops a putrid odour, becoming a 'stinker'.

Predisposing factors include; low moisture content of compost (less than 64 per cent) coupled with high moisture in casing soil.

Brown Discolouration

Browning of small pin heads or half grown mushroom is very common in seasonal mushroom growing farms. This may be caused by high temperature, spraying at higher water pressure, with too high chlorine rate or incorrect use of formalin *e.g.* by spraying the mushrooms with a formalin solution.

As compared to white button mushroom, there are few physiological disorders, recorded in oyster mushroom. Reduced light in the cropping rooms, result in longer and thicker stipe and pileus is partly reduced. Insufficient ventilation and low light exposure, induce bunched growth regenerate.

Chapter 9

Post Harvest Handling and Preservation of Mushrooms

Diversified agro-climatic conditions in India offer vast potential for growing different types of mushrooms. There are about 30 genera of mushroom being cultivated throughout the world as food, only four *viz.*, white button mushroom (*A. bisporus*), oyster mushroom (*Pleurotus* spp.), paddy straw mushroom (*Volvariella volvacea*) and milky mushroom (*Calocybe indica*) are grown commercially in India. Out of these, white button mushroom contributes about 85 per cent of the total 'Mushroom production'. Mushrooms are essentially a rich source of good quality proteins, minerals and vitamins but being low in calorific values can also be used to supplement the diet of a diabetic patient. Lately, it has been observed that mushroom contain certain anti-carcinogenic substances which are being used for cancer therapy in some advanced countries.

In India, mushroom cultivation has gained much importance owing to its high economic returns. Despite increase in the production of mushrooms, more than 70–75 per cent of the annual production reaches in limited markets within short duration of 2-3 months being highly perishable because of their high moisture content and delicate nature. Due to short life, like fruit and vegetables/mushrooms are affected by many post harvest changes like wilting, browning, liquefaction, loss of texture, aroma, flavour etc., creating thereby market problems. These problems have consequently, led to the development of some post harvest preservation techniques to enhance their shelf life without any appreciable change in nutritional quality. Therefore, understanding of post harvest handling practices plays a significant role in enhancing the availability of quality mushrooms either in fresh or processed form to the consumers and at the same time ensuring remunerative prices to the producers.

Processing and Preservation

As far as processing technologies are concerned, sun drying of the mushrooms is one of the simple and oldest method followed by growers, since time immemorial. Due to the difficulties in drying of some of the mushrooms, new preservation technologies like cabinet drying, canning, pickling freeze-drying and by irradiation treatment mushrooms have developed to improve the shelf life and consumption of mushrooms.

In the peak period of harvesting, the gluts in the market can be checked by adopting appropriate post harvest technologies. Development of appropriate storage and processing technology in order to extend their marketability and availability to consumer in fresh as well as processed form is of great significance. The post harvest technology includes (1) Harvesting, (2) Pre cooling, (3) Sorting (4) Dips treatments to extend storage life (5) Packaging (6) Transportation (7) Storage (8) Processing (9) Marketing and export.

Many short term storage measures are followed to retard the deterioration in quality at the level of mushroom grower till it reaches the consumer. Short term preservation methods like pre-packaging coupled with low temperature storage, irradiation and steeping preservation help to prolong storage life from one to three weeks. Prolongation of stalks, opening of veils, toughening of texture and loss of whiteness are some of the deteriorative changes associated with storage. Long term preservation methods, such as canning, drying, pickling etc. can make the availability of mushroom of good quality throughout the year at reasonable cost. Freezing and freeze drying methods have also been recommended for long term storage but these are energy intensive, sophisticated and cost prohibitive techniques to be used for export. Mushroom can also be successfully utilized in the weaning foods, biscuits, soup powders, and in a number of dishes in the kitchen. Various aspects of post harvest technologies in mushroom are given as under:

Short Term Storage

Post harvest management of fresh mushroom:

Harvesting

Mushrooms should be harvested at the optimum stage of maturity. Harvesting of under mature or over mature fruit bodies results in poor texture, flavour and immediate degradation.

White button mushrooms should be harvested when cap diameter (30-45 mm) is twice the length of stem. Pre-harvest spray of 2 per cent ascorbic acid improves the colour by inhibiting the polyphenol oxidase activity.

Oyster or dhingri mushrooms are harvested when the fruit bodies has curled under edges and with well formed gills. As the maturity advances, the fruit body loses water and thus causing shrinkage and deterioration of quality.

Paddy straw mushroom should be harvested at button or egg stage. The deterioration is less in the mushrooms harvested at egg stage because the polyphenol oxidase activity is slow. Due to fast post harvest maturity, many changes occur as the cap open and it gets the appearance of flattened umbrella.

Pre-cooling

The temperature of the button mushroom after picking varies between 15°C to18°C and it rises steadily during storage due to respiration and atmospheric, temperature. They are, therefore, required to be pre-cooled immediately after harvest to their optimum storage temperature (5°C) to check the rate of high respiration and deteriorative changes which are otherwise very fast at ambient room temperatures. Vacuum cooling and ice-bank cooling of mushroom is now in vogue.

Sorting

While sorting out undesirable fruiting bodies various attributes are taken into consideration such as discolouration, blemishes and shape.

Dips/Treatments

Because of the unavailability of peat as casing material in India, washing becomes necessary to remove soil particles. The washed mushrooms generally deteriorate rapidly than mushroom packed dry. Fresh white mushroom are white or light buff without dark marks on either cap or stem, the veil is closed and the gills are not visible. To maintain whiteness, dipping of mushroom in dilute solution of hydrogen peroxide (1:3) for half an hour and then steeping in 0.25 per cent solution containing 500ppm sulphur dioxide has significant affect. However, in order to remove any adhering casing soil, compost or other substrates and to maintain whiteness, the produce is washed in 0.025 per cent to 0.05 per cent potassium metabisulphite solution.

Packaging

Properly washed/treated with chemical solution, the mushrooms are packed in suitable containers with the main objectives of proper protection and presentation of produce, thereby enhancing the shelf life and thus extending the market period. While a good package sells a product, a mediocre package can interfere with sale of an otherwise excellent product. Since mushrooms are very sensitive to desiccation and draught, selection of suitable package is very important. Mushrooms are usually packed in polythene bags of 200–500g capacities. Mushrooms packed in more than this capacity bags have a tendency to loose acceptability. In 100 gauze polyethylene bags 0.5 per cent ventilation is generally recommended for refrigerated storage, whereas, for local markets non-perforated bags should be used. For long distance markets polystyrene or fiber board punnets over wrapped with partially permeable polyvinyl chloride or polyacetate films packing material is recommended. These over wrapping create modified atmosphere in the package, producing an atmosphere of about 10 per cent CO_2 and 2 per cent O_2.

Transportation

For carrying the mushrooms for long distances markets, the transportation should be done in refrigerated vans. Pre-cooling of produce is must before loading into refrigerated vans. For local markets, the pre-cooled mushroom packed in suitable packages should be transported in insulated ice-containers to check the rate of respiration and deterioration in white colour.

Storage

Generally mushrooms can not be stored more than 24 hours at room temperature or 1-2 weeks in refrigerated atmosphere. During storage there is decline in total sugars, soluble proteins and total phenol content, while polyphenol oxidase activity enhances. These changes are aggravated at high storage temperature. A temperature of 5°C alongwith 85-90 per cent relative humidity is generally recommended for the storage of mushrooms for better retention of vitamin C and other nutrients. Pre-storage chemical treatments with salt, potassium metabisulphite, citric acid and tartaric acid individually or in combination result in reducing discolouration during storage at room or cool temperatures. At ambient temperature, the steeped mushrooms can be kept for 8-10 days. However *Pleurotus sajor-caju* could be kept upto 3 months in a steeping solution containing 5 per cent salt, 0.2 per cent acetic acid and 0.1 per cent potassium metabisulphite in glass bottles without loosing much of its organoleptic quality. Irradiation with gamma rays (1-2 KGy dose) has been found to increase the storage life for 9-10 days at 3-5°C and 2-3 days at ambient temperature. The best way of storage of paddy straw mushroom is at 10-15°C in polythene bags with perforation.

Long Term Storage

Processing Technologies (White Button Mushroom)

Freshly harvested mushrooms are highly perishable and require immediate processing for stable preservation. Pickling and sun drying are economically viable methods of preserving mushrooms during the glut period. However, freezing, freeze drying and canning gives a product of excellent quality and can be explored commercially for export market. The following methods have potential for processing mushroom into different products for long term storage:

Preservation by Low Temperature (Freezing)

Individual quick freezing technology is employed to yield a product of excellent quality with longer shelf life. Quick freezing is a process, where temperature of the mushroom passes through the zone of maximum crystal (0-3°C) in 30 minutes or less. To control the activity of enzyme, mushrooms prior to freezing are required to be blanched. For best results, storage temperature should be -35 to 40°C.

Preservation by High Temperature (Canning)

The main principle of preservation by the use of high temperature is to destroy the micro organism and to prevent the re-contamination by hermetic sealing. This is a well practiced technique all over the world. During recent years, retortable glass bottles and flexible pouches are also being used due to high cost of cans. Generally for canning purpose, small sized buttons are required without stem attached. Ascorbic acid, ethylene diamine tetra acetic acid and citric acid have been recommended as useful adjunct for improvement of colour in canning of mushrooms. Steeping the buttons in 0.5 per cent solution of methyl cellulose or carboxymethyl cellulose prior to canning result in increasing the yield by 5.9 per cent without any deleterious effect on quality. Food additives like agar (0.125 per cent) and pectin (0.5 per cent) improves the flavour of canned products. The canning process can be divided into six operations, namely, Cleaning, blanching, filling, sterilization, cooling, labeling and packaging. The storage before canning is done at 4-5°C. The mushrooms without stem length of 0.5 to 1 cm are preferable for canning and these are canned whole or sliced. The methodology is given as under:

Methodology

White button mushroom (*Agaricus bisporus*) is preferred over other types of mushrooms for canning. Commercially, mushrooms are canned in brine. The various steps are as follows:

Picking

Mushrooms are picked at button stage (cap diameter 2-2.5 cm). The soil and portion carrying any microbial flora is cut off. The stalk length should preferably be kept 0.5-1 cm long.

Sorting and Grading

Healthy, white and tight buttons are selected and separated into two grades *i.e.* cap diameter upto 2.5 cm, with compost head as "A" and cap diameter beyond 2.5 cm as "B" grade.

Washing

Graded mushrooms are thoroughly washed 3.4 times in cold running water to remove adhering, dirt, soil etc.

Blanching

Mushrooms are blanched in boiling water for 2-3 minutes followed by immediate cooling in cold water.

Filling of Cans

Mushrooms are commercially packed in two can sizes *i.e.* A-1, tall can and A 2½ cans. The blanched mushrooms are filled into cans to its declared drain weight *i.e.* 230g in A 1½ tall can and 430g in A-2 can size.

Brining

Strained hot brine solution having a ratio of 2 per cent common salt, 1 per cent sugar and 0.05 per cent citric acid is added upto the brim of the can.

Exhausting

After brining, cans are exhausted to remove any entrapped air and other occluded gases from the product to ensure a longer shelf life. Exhausting can also be performed by placing the filled cans in boiling water till temperature of the centre of the can reaches 85-90°C for 1-2 minutes.

Seaming/Can closing

Immediately after exhausting, cans are sealed with the help of a double seamer to get hermetic seam. Sealed cans are then placed in upside down position to sterilize the coded lids.

Processing/Sterilization

This is accomplished by processing the hermetically sealed cans at a pressure of 15 lb psi for a specified period of time (usually 35-45 minutes).

Cooling

Cooling of the can is done immediately after sterilization in cold running water to room temperature.

Labeling and Storage

The cooled cans are stored in a cool dry place and smeared with grease to remove any adhering moisture from the can body to avoid rusting. Cans are kept at ambient temperature for 8-10 days to check any swell, leakage, puffing and other disorders before labeling.

Before, the cans are exposed for sale, proper labeling is done to meet statutory requirement of Fruit Products Order.

Preservation by Drying

The main objective of drying is the removal of free water to such a level (aw< 0.7) that the biochemical and microbial activity is checked due to reduced water activity. The removal of water from substance by direct sublimation from frozen state to vapour state is known as freeze-drying. In a freeze-drying system the use and shape of original product can be retained and shrinkage is negligible. Freeze drying yields a product of excellent quality but the cost of removal of water is ten times higher than the conventional air drying. Sun drying is the appropriate technology for our country, because of free source of energy and minimum investment in capital cost. Blanching and sulphuring/sulphiting are important treatment given before drying. Pre-drying soaking of blanched slices in potassium metabisulphite (1 per cent) + Citric acid (0.2 per cent) + Sugar (6 per cent) + Salt (3.0 per cent) solution for 16 hours followed by drying at 60±2°C for 8.5 hours yields best product with longer shelf life.

Preservation by Chemicals

There are several chemicals additives which are non-nutritive and are added in small quantities in foods to improve its appearance, flavour, texture and storage properties. Mainly used additives are salt, sugar, acetic acid, vinegar, spices, essential oil etc. On the other hand, chemical preservatives maintain nutritional quality and enhance the keeping quality. Potassium metabisulphite is commonly used preservative in mushroom processing.

Preservation by Lactic Acid Fermentation

Fermentation is a process of anaerobic or partial anaerobic oxidation of carbohydrates. During the process of fermentation, sufficient quantity of lactic acid is produced to prevent the product from further spoilage during storage.

Preservation by Irradiation

Application of gamma radiation retards the deteriorative processes and help in extending the shelf life of mushrooms. Doses of 1-2 KGy delays the cap opening and stem elongation in button mushrooms and thus extend its marketability.

Preservation by Minimal Processing

It has been observed that if different processes/chemical preparation are combined in judicious manner, excellent product with better retention of original colour, flavour and nutrition can be obtained.

Preservation by Pickling

One of the most economically viable methods of preserving mushrooms for longer period is by way of pickling. Mushroom are washed sliced and blanched for 5 minutes in 2 per cent brine + 0.05 per cent potassium metabisulphite solution. The blanched mushrooms are washed in cold water for 2-3 times and the excess water is drained off. To these mushroom 10 per cent sodium chloride is added and kept over night. The excess water is drained off on the next day and appropriate preservatives and spices are mixed to get organoleptically acceptable pickle. Acetic acid or sodium benzoate is added as preservatives. Turmeric, rye, red chillies, cumin, aniseed, black pepper powder, jagari, ajwain, kalongi, alongwith mustard oil are added to the processed mushrooms. The pickle, thus prepared can be stored in plastic jar for more than one year.

Oyster Mushroom (*Pleurotus* spp.)

For preservation of oyster mushroom for long time dehydration is a well established and most commonly used technique. Sun drying is carried out by spreading the fruit bodies on polythene or newspaper sheet, under the sun, when the temperature and humidity ranges between 25-30°C and 40-45 per cent respectively for at least two days. Sun dried product should be dried at 50-52°C in oven for 5 hours to avoid further spoilage.

Paddy Straw Mushroom (*Volvariella* species)

Sun drying of paddy straw mushroom is the common practices of long term storage. Before drying the fruit bodies are blanched and treated with chemicals. Best dried products in respect of colour, texture and redehydration properties were obtained when the fruit bodies were steeped either in 1 per cent potassium metabisulphite solution alongwith 0.2 per cent citric acid or in 0.5–1.0 per cent of citric acid alone.

Chapter 10
Mushroom Poisoning

From the earliest days of mankind, edible mushroom have been appreciated for their excellent culinary characteristics and gastronomical properties, and even today in most countries, a bone cuisine without mushroom is unthinkable. There are about 30 genera, splitted into 80 species and more or less 500 cultivated strains/forms which are presently being domesticated either on a small or large scale. It has been estimated that out of 10,000 species of fleshy fungi about half of them are edible and as many as 100 species are poisonous and therefore, consumption of wild mushroom require adequate precautions to avoid accidental death due to mushroom poisoning. Though many species are known to be edible, the production of cultivated mushroom is still dominated by *Agaricus bisporus* which accounts for approximately 32 per cent of the world's total production. The next important species in terms of world output is *Lentinula edodes* followed by species of *Pleurotus, Auricularia* and *Volvariella.*

Poisonous Mushrooms

The term 'Mushroom' in its broad sense include all fleshy fungi which are edible. However, the word mushroom has no significance in distinguishing the edible from poisonous one. The term 'Toadstool' includes all those fleshy fungi which are popularly looked upon as poisonous to eat.

The distrust and dread which most people entertain towards the eating of fungi are without doubt due to fear of committing an error that may prove fatal. This fear is mainly due to lack of knowledge and understanding. In collecting fungi, for table purpose, an open eye should be kept while going through the forests, fields, roadsides or mountain sides, and only those fungi should be picked up whose edibility is well recorded and with which the collector is also familiar. Only healthy specimens should be collected, discarding all over mature and decayed specimens. Eating of under developed specimens should also be avoided. With adequate knowledge and familiarity with these fungi and with due care during collection of fungi, trouble due to eating of poisonous kinds can be avoided fully. In case of poisoning, the doctor should be informed of the suspected cause of trouble without delay and without exception. As a first aid measure, stomach and intestine should be emptied as promptly as possible. Vomiting should be induced without delay. After vomiting, give plenty of lukewarm water to the patient and keep him warm and provide heat locally to relieve pain.

In case, one exercise due care and possesses adequate knowledge about distinguishing edible and non-edible kinds, one will enjoy eating of this vast group of edible fungi.

Nature of Mushroom Poisoning

There are some mushrooms that contain poison that are inactive when taken orally (*e.g.* bufotenin in *Amanita citrina*) or are in less amount to be physiologically active (*e.g.* HCN in *Marasmius oreades*). Some mushroom including common *Agaricus bisporus* can cause digestive upsets in some people but they cannot be called poisonous because most people eat them without adverse effects. Nature of mushroom poisoning can be summarized as below:

1. Some mushrooms are poisonous raw, but harmless when cooked; *e.g.* the (wood) Blewit, *Clitocybe nuda.*
2. Some mushrooms are poisonous unless paraboiled and then thoroughly cooked; *e.g.* the False morels, *Gyromitra esculenta.*

3. Some mushrooms are poisonous regardless of the manner of their preparation *e.g.* the Death Cap, *Amanita phalloides*.
4. Some mushrooms are poisonous only under certain conditions, such as when consumed alongwith alcoholic beverages, *e.g.* the Inky Cap, *Coprinus atramentarium*.
5. Some mushrooms are poisonous only when eaten in large quantity, *e.g.* the Early Morel, *Verpa bohemica*.
6. Some mushrooms are classified as poisonous, but since the characteristics symptom is hallucinogenic intoxication, and since these mushrooms are generally eaten to achieve this effect, they are not considered poisonous by the user except in case of overdoses or unexpected adverse reactions, *e.g.* the Liberty Caps, *Psilocybe semilanceata*.
7. Some mushrooms, for reason unknown are deadly poisonous in some geographical areas but edible in others; *e.g. Paxillus involutus*.
8. Some mushrooms become poisonous (*i.e.* can cause stomach upset) when they are old, decayed or hit by frost, although they are edible when fresh and young *e.g.* the Honey Mushroom, *Armilearia mellea*.

Types of Mushroom Poisoning

Five types of mushroom poisoning have been recognized which are as follows:

Gastero-intestinal Type

Characterized by early appearing of symptoms of nausea, vomiting and diarrhoea. The symptoms terminate rapidly and usually spontaneously the patient being restored to health in a day or two. Causative mushrooms are *Russula emetica*, *Lactarius torminosus* and *Entoloma lividum*.

Choleriform Type

Gastro intestinal symptoms develop in10 to 15 hours, followed by rapid loss of strength and weight. The death rate is high. Causative mushrooms are *Amanita phalloides* and *A. virosa*.

Nerve-affecting Type

Early gastro–intestinal symptoms appear in 2-3 hour after eating and terminate into violent convulsions, coma and often death. The active principle involved is 'muscarin' for which the perfect antidote is 'atropin'. Causative mushroom are *Amanita muscaria, Inocybe infelix* and *Clitocybe illudens*.

Blood Dissolving Type

Abdominal stress with jaundice developing in 4 to 5 days, death may occur in very rare cases. Blood transfusion has been suggested by Ford as the logical treatment causative mushroom is *Gyromitra esculenta*.

Cerebral Type

The symptoms appear shortly after consuming the mushroom. The symptoms are exhilaration, staggering gait, and queer disturbance of vision. The patients are soon normal again. Causative mushrooms are various species of genus *Panaeolus*.

List of fungi which are poisonous to eat are given in Table 4.

Table 4: Fungi which are Poisonous to Eat

Sl. No.	Scientific Name	Common Name	Poisonous Substance	Poisoning Potential	Type of Poisoning	Identification	Habitat
1.	*Amanita phalloides*	Death angel	Two toxic principles have been characterized. One is 'Amanita hemolysin' which is heat labile and other is *Amanita* toxin which is a complex of L-amanite ß-amanitine and Phalloidin	Deadly poisonous	Choleriform	Vary in colour, in the manner in which the volva is ruptured and the surface of items. Presence of volva and ring is consistant feature of the genus	It occurs commonly in wood, groves, mixed conifers and hardwood from July to October
2.	*A. virosa*	Destroying angel	–do–	Deadly poisonous	Choleriform	It is usually much larger than others. Very much like *A.verna* but differ in elevated almost unbonate center of cap and in the free conditions of the volva	–do–

Contd...

Table 4–Contd...

Sl. No.	*Scientific Name*	*Common Name*	*Poisonous Substance*	*Poisoning Potential*	*Type of Poisoning*	*Identification*	*Habitat*
3.	*A. verna*	Destroying angel	–do–	Deadly poisonous	Choleriform	This is a pure white large species with prominent death cap enclosing the base of stalk very closely.	It occurs in hard wood and mixed forests during wet season (July-September)
4.	*A. muscaria*	Fly agaric	Muscarin and Mycoatropine	Poisonous	Nerve affecting	Easily identified on account of its scaly cap, often brilliantly coloured, its large ring and volva on its bulb like base.	It occurs on road sides rocky mountain and open forests of conifers and hard wood (July-Sept.)
5.	*Amanita pantherina*	Panther Amanita	Muscarin and Mycoatropine	Poisonous	Nerve affecting	The warts on the cap are well formed and whitish to cream. The narrow, often in rolled collar like volva are distinguishing characters.	It grows generally scattered to gregarious under conifers and hard wood from June to October

Contd...

Table 4–Contd...

Sl. No.	Scientific Name	Common Name	Poisonous Substance	Poisoning Potential	Type of Poisoning	Identification	Habitat
6.	*Clitocybe illudens*	Jack-My-Lantern	Muscarin type	Mildly poisonous	Nerve affecting	The clustered habit of growth, yellow orange colour of fruit bodies crowded, narrow, decurrent gills and stalk tapered to the base. Fresh fruit body emits light in darkness.	It grows from July to October in large cluster at the base of decaying stem.
7.	*Inocybe infida* and *I. infelix*	Unfortunate inocybe	Muscarin	Slightly dangerous	Nerve affecting	In mature state the cap is innately silky or fibrillose scaly.	It grows from May to August in mossy grounds.
8.	*Lactarius torminorus*	Milky Cap	Not characterized but substance is easily destroyed by heat	Mildly poisonous	Gastero interstinal	This species has an inviting appearance, coral pink tint, woody margin, zoned surface and peppery taste.	It is found growing in woody forests or on rotton logs in wet summor from July to October.

Contd...

Table 4–Contd...

Sl. No.	Scientific Name	Common Name	Poisonous Substance	Poisoning Potential	Type of Poisoning	Identification	Habitat
9.	*Russula emetica*	Pungent Russula	Muscarin	Slightly poisonous	Gastero-intestinal	The fungus can easily be recognized at maturity by its slate green gills. Volva is lacking.	It is found growing in woody forests or on rotton logs in wet summer from July to October.
10.	*Lepiota molybdites*	—	Not characterized	Slightly poisonous	Gastero intestinal	The cuticle peals off readily from the cap exposing the faint red surface of flesh immediately underneath it.	It is found growing in pasteurs and grassy areas common during June to Sept., and often forms large circles.
11.	*Gyromitra esculenta*	False Morel	Melvellic acid	Contradictory	Blood dissolving	The cap is not regularly pitted as in case of Morchella with which it is oftenly confused. The cap is less irregular like the convolutions of the bran.	It is abundant in the coniferous forests and fungus fruits in spring.

Contd...

Table 4–Contd...

Sl. No.	Scientific Name	Common Name	Poisonous Substance	Poisoning Potential	Type of Poisoning	Identification	Habitat
12.	*Panaeolus companulatus*	Black-spored mushroom	Not characterized	Slightly poisonous	Cerebral	Spotted gills and long narrow stems are important characteristics of the group.	Common in dung or rich grassy manured grounds from May to October.
13.	*Psilocybe* species	Psychotrophic fungi	Psilocybin and Psilocin	Some times deadly	Cerebral	The margin of the cap in fresh specimens is occasionally striatulate, cap broad, thin, companulate or convex, glaborous brown to redish brown when moist, paler when dry	Common in tufts, among fallen leaves or on and about the base of tree from June to October.

Bibliography

Arrold, N.P. and C.D.Blake, 1966. *Nematologica* 12: 501-510.

Atkin, F.C. 1981. *Mushroom Growing Today*. Faber and Faber, London pp.188.

Atkin, F.C. and C.L.La Jouche. 1948. *Mushroom Dis.Leaft* 3:3

Azad, K.C., M.P.Srivastava, R.C.Singh and P.C.Sharma. 1987. *Ind.J.Mush*. XII- XIII: 21-29.

Bajaj, H.K. and K.K.Walia. 1999. *Ind.J.Nematol*. 29(2): 154-170.

Bano, Z. and H.C.Srivastava. 1962. *Food Sci*. 12: 363-65.

Bartan, R.J. 1985. *Mush.J*. 151: 223-228.

Bech, K. and C.R.Rasmussen. 1969. *Mush.Sci*. 8:329-42.

Beelman, R.C.; B.D.Guthril and D.J.Royse. 1989.; *Mush.Sci*. 12(Part.II): 655-665.

Beyer, D.M. 1994. *Mushroom News* 7: 7-20.

Bhandal, M.S. and K.B.Mehta. 1989. *Mush. Sci*. 12(I): 25-35.

Bhatt, N. and R.P.Singh. 2000. *Mush.Sci*. 15: 587-593.

Block, S.S., G.Tsano and L.Han. 1958. *Mush.Sci*. 4: 309-25.

Bose, S.R. 1953. *Indian Pharm.* 15: 279-281.

Brandl, F. 1994. *Champignon* No.381, 202-206.

Cailleux, K. 1969. *J.Gen. Microbiology* 105: 135-151.

Cantwell, G.E. and W.W.Cantelo. 1994. *J.Econ.Ento.* 77(2): 474-75.

Chadha, K.L. 1994. *Mushroom Research.* 3(1): 1-4.

Chang, S.T. 1979. *Mush.Sci.* 10: 609-618.

Chang, S.T. 1990 b. *Mush.J.*295: 422-423.

Chang, S.T. 1991. *Mush.J.Tropics* 11: 42-52.

Chang, S.T. 1999. *J.Med.Mush.* I: 291-300.

Chang, S.T. and P.G.Miles. 1989. *Edible Mushroom and Their Cultivation.* Boca Ratton Florida, U.S.A. CRC Press Inc.

Chang, S.T. and P.G.Miles. 1991. *Mush.J.* 504: 15-18.

Chen, K.P., M.YuLiang and S.T.Chang. 1988. *Mycologia* 80: 849-853.

Cheng, S.C. and C.C.Tu. 1978. *The Biology and Cultivation of Edible Mushroom* (S.T.Chang and W.A.Hayes eds) pp.605-625. Academic Press. New York.

Clift, A.D. and M.A.Jerras. 1992. *Plant Protection Quarterly* 7(2): 59-61.

Constantin, J. and L.Matruchot. 1894. *Rev.Mycology* 11: 62-66.

Chibata, I., K.Okumura, S.Tekeyama and K.Kotera. 1969. *Experientia* 25: 1237-1238.

Chihara, G., J.Hamuro, Y.Y.Maeda, Y.Arai and F.Fukuoka. 1970. *Cancer Research* 30: 27-81.

Crisan, E.V. and Anne Sands. 1978. *The Biology and Cultivation of Edible Mushroom* (S.T.Chang and W.A.Hayes eds.) pp.137-168. Academic Press, New York.

Cross, M.J. and L.Jacobs. 1968. *MGA Bulletin* 238: 440-447.

Culpeper, N. 1952. Complete Herbal. Foulshan. 1952. Reprints in London.

Das, L. 2003. *Current Vistas in Mushroom Biology and Production* (R.C.Upadhyay, S.K.Singh and R.D.Rai eds.) pp.253-256. Mushroom Society of India, NRCM, Chambaghat (H.P.).

Das, L. and M.Suharban. 1991. *Indian Mushroom* (M.C.Nair eds.) pp.30-45. Kerala Agricultural University Vellar.ikkara.

Delmas, J. 1978. *The Biology and Cultivation of Edible Mushroom* (S.T.Chang and W.A.Hayes eds.) pp.251-299. Academic Press, New York.

Deritex, P.J.L., F.H.M.Simons, H.J.M.Op.Den Comp, Venller Dlift, L.J.L.D.Van Griensven and G.D.Vogels. 1991. *Environmental Microbiology* 57: 563-567.

Dieleman-Van Zaayen, A.1970. *M.G.A.Bulletin*. 244: 158-178.

Dhar, B.L. 1997. *Advances in Mushroom Biology and Production*(R.D.Rai, B.L.Dhar and R.N.Verma, eds) pp.369-378. MSI, NRCM, Chambaghat (H.P.)

Dhar, B.L. 2002. *Recent Advances in the Cultivation Technology of Edible Mushroom* (R.N.Verma and B.Vijay eds.) pp.57-69. MSI-NRCM, Chambaghat(HP).

Dmosaska, E., J.Omoch and K.Ilieva. 1977. *International J.Mush.* 2(1): 15-23.

Doshi, A., S.S.Sharma and A.Trivedi. 1991. *Advances Mush.Sci.* page 57(Abst.)

Doshi, A. S.S.Sharma and A.Trivedi. 1993. *Mushroom Information* 10: 14-22.

Doshi, A. and R.D.Singh. 1985. *Ind.J.Mycol.Pl.Pathol.* 15: 269-273.

Duggar, B.M. 1905. Bull.85. Bur.Plant Industry. U.S.Deptt. of Agriculture.

Dull, G.G. 1978. *J.Food Prot.* 41: 50-53.

Edward, R.L. 1950. *Mush.Sci.* 1: 62-63.

Eicker, A. 1987. *South African J. Botany.* 53(5): 345-348.

Elliot, T.J. 1985. *The Biology and Technology of Cultivated Mushroom* (P.B.Flegg, D.M.Spencer and D.A.Wood eds.) pp.111-139. Wiley.

Elliot, T.J. 1986. *Mush. J.* 125: 69.

Elliot, T.J. and M.P.Challen, 1974. Annual Report Glasshouse Crop Res.Institue, Littlehempton pp.194-204.

Fahy, P.C. 1981. *Mush.Sci.* 71: 293-311.

Falck, R. 1923. *Chem.Ber.* 56B: 2555-2556.

Fang, Y.S. and G.J.Luo. 1991. *Jour.Human Ag.College* 19: 39-46.

Fermor, T.R. 1986. *Cultivating Edible Fungi* (P.G.Wuest; D.J.Royse and R.B.Beeleman eds) pp.361-370. Elsevier Science Publishers, The Netherlands.

Fletcher, J.T. 1975. *Mush.J.* 25: 4-5.

Fletcher, J.T., P.F.White and R.H.Gaze. 1989. *Mushroom: Pest and Disease Control* 2nd Edition. Intercept, England, 174pp.

Gandy, D.G. and M.Hollings. 1962. *Rep.Glasshouse Crop Res.Inst.* 1961: 103-108.

Ganeshan, G. 1987. *Mushroom Information* 4: 12-13.

Garcha, H.S. and A.Sekhon. 1981. *Mush.Sci.XI(1):* 411-417.

Garcha, H.S., P.K.Khanna and G.S.Sandhu. 1987. *Cultivating Edible Fungi* (P.G.Wuest, D.J.Royse and R.B.Beelman eds.) pp.325-332. Elsevier Science Publishers, The Netherlands.

Giacomini, V. 1957. *Sci.Aliment* 3: 103-108.

Gillespie, D.R. and D.M.J.Quising 1990. *Can.Ent.* 122: 975-983.

Goulas, J.P. 1994. *Mushroom Information* 2/8: 5-21.

Gray, W.D. 1973. *The use of Fungi as Food and in Food Processing* (Part-II) pp.81 CRC, Press Ohio.

Greuning, M., A.Eicker and G.M.Van. 1992. *Mushroom Research* (1): 33-38.

Grewal, P.S. and C.Smith. 1995. *Mushroom News* 43: 15-25.

Griensven, L.J.L.D. 1988. *The Cultivation of Mushroom* (Publ.Mushroom Expt.Station, Horst, Holland).

Halachmy, I. and C.H.Mannheim. 1991. *Packaging Technology and Science.* 4: 279- 286.

Harmsen, M.C., B.Joulner, A.Kram, S.J.Go, De Haan and J.g.H.Wessels. 1991. *Curr.Ganet.* 20: 137-144.

Harper, E., F.C.Moler and B.J.Macauley. 1992. *Australian Journal of Experimental Agriculture* 32: 657-667.

Hawksworth, D.L. 1991. *Mycological Research* 95: 64-65.

Hayes, W.A. 1992. *Mush. J.* 507: 16-17.

Hayes, W.A. and T.R.Shandilya. 1977. *Ind. J. Mycol. Pl. Pathol.* 7: 5-10.

Hollings, M. 1978. *Adv. Virus Res.* 22: 1-53.

Hollings, M., D.G.Gandy and F.T.Last. 1963. *Endeavour* 22: 112-117.

Holtz, R.B. and C.L.Schirler. 1971. *Lipids* 6: 176-180.

Hseu, R.Y. 1993. An overview on *Ganoderma* Mushroom Jai Chung, Taiwan, Wann Nain Publishing Co. pp.151.

Hsu, Haang-Kao and Han You-Hsin. 1981. *Mush.Sci.* 11: 403-425.

Hussey, N.W., W.H.Read and J.J.Hesling. 1969. *The Pests and Protected Cultivation*. London Arnold.

Ikekawa, J., N.Uchara, M.Nakemishi and F.Fukuoka. 1972. *Cancer Res.* 29: 734-735.

Ingratta, F.J. and H.A. Olthof.1979. *Mush.Sci.* 10: 394-405.

Iracabal, B., G.Zerkakis and J.Labarere. 1995. *Microbiology* (Reading). 141: 1479-1490.

Jacob, E.H.1922. United State Patent I, 434-822.

Jandaik, C.L. and J.N.Kapoor. 1974. *Mush.Sci.* 9(2): 667-672.

Jandaik, C.L., V.P.Sharma and Rajeev Raina. 1993. *Mush.Res.* 2: 45-48.

Jhune, C.S., G.P.Kim and D.Y.Cma. 1991. *Korean Journal Mycology* 19: 85-90.

Kachroo, J.L. 1991. *Ind.J.Mycol.Plant Pathol.* 21: 287-288.

Kannaiyan, S. and N.N.Parsad. 1978. *Indian Mush.Sci.* 1: 287-293.

Kaufman, T.D., J.R.Bloom and F.L.Lukezic. 1983. *Mush.News*, 31: 9-10.

Kaul, T.N., J.L.Kachroo and N.Ahmed. 1978. *Indian Mush.Sci.*I:193-203.

Kaul, T.N. and K.K.Janardhananan. 1970. *Ind.Phytopath.* 25: 578-580.

Khanna, A.S. 1993. *Curr.Nematol.* 4: 173-176.

Khanna, A.S. and N.K.Sharma. 1989. *Nematol. Mediterr.* 17: 61.

Khanna, A.S. and N.K.Sharma. 2000. *Ind.J.Mush.* 18(1 and 2): 60-61.

Kim, K.C. and C.Y.Hwang. 1996. *Korea J.Appl.Ento.* 35(1): 45-51.

King, T.A. 1993. *Mushroom News* 41(2): 29-46.

Kissop, B.E. and A.Doyle. 1991. *Maintenance of Micro-organism and Cultivated Cells*. 2nd ed. Academin Press, London.

Kumar, S. and S.R.Sharma. 1998. *Mushroom Research* 7(1): 25-28.

Laborde, J. 1993. *Mushroom Information*. 9: 5-27.

Laborde, J. 1994. *Mushroom Information*. 12: 5-15.

Lal, B.B. and K.D.Sharma. 1997. *Advances in Mushroom Biology and Production* (R.D.Rai, B.L.Dhar and R.N.Verma, eds.) pp.223-273. MSI, NRCM, Chambaghat (HP)

Lambert, E.B. 1941. *J.Agric.Res*. 62: 415-422.

Last, F.T. 1970. *M.G.A.Bulletin*: 259-273.

Lamke, P.A. 1976. *Ann.Rev.Microbiol*. 30: 105-145.

Li, G.S.F. 1982. *Mush.Sci*. 10(2): 609-618.

Lind, R. 1993. *Contract Res.Hort.Dev.Council*. 28.

Liu, F., Vec Ooi and S.T.Chang. 1995. *World.J.Microbiol.Biotechnol*. 11: 486-490.

Liu, H.P. and P.Liang. 1986. *Acta Microbiol. Sinica* 26: 221-225.

Lundberg, G. 1944. *Smb.Bot.Ups*. 8: 1-183.

Malthouse, G.T. 1901. *Microsc.Soc.Part(III)*: 182.

Mamoun, M. and J.M. Olivier. 1995. *Mush.Sci*.14: 659-677.

Mantel, E.F.K., 1973. *Ind.J.Mush*. I: 15-16.

Mantel, E.f.K., R.K.Aggarwala and P.K.Seth. 1972. *A Guide to Mushroom Cultivation*. Min. of Agri., Farm Information Unit Directorate of Extension, New Delhi Farm Bull. 2.

Maul, S.B., P.A.Lemke, W.L.Derner and J.B.Yodder. 1980. United State Patent 4: 204- 384.

Miller, F.C. 1994. *Agaricus* compost (N.G.Nair eds.) Proc.of the 2nd AMGA/ISMS International Workshop-Seminar on *Agaricus* compost. Sydney, Australia: 1-18.

Miller, F.C. and M.Spear. 1995. *Mush.Sci*. 14(2): 635-642.

Morris, E., Doyle, O. and Claney, K.J. 1995. *Mush.Sci*. 14(2): 619-625.

Nair, N.G. 1973. *Aust.J.Agaric.Res*. 24: 533-541.

Nair, N.G. 1977. *Plant Disease Survey* (1975-1976): 30-32.

Namba, H. 1993. *Mushroom News*. 41(2): 22-25.

Nath, A., S.K.Patyal and V.P.Sharma. *Current Sci.* 76: 870.

Nielsen, O. 1932. Reprinted from Gartnertideake: 2pp.

Oliff, K.E. 1965. *Hort.Res.* 5; 4.

Paine, S.G. 1919. *Ann. Appl. Biol.* 5: 206-209.

Passmore, E.L. and R.R.Frost. 1979. *Phytophathology*. Z.95: 346-350.

Pizer, N.H. 1973. *J.Agric.Sci.* 27: 349.

Poppe, J.A. 1972. *Mush.Sci.* 8: 517-525.

Poppe, J.A. 1978. *Mush.Sci.* 10: 851-872.

Preece, T.F. and W.G.Wang. 1982. *Physiol.Plant Pathol.* 21: 251-257.

Purkayastha, R.P. and A.K.Das. 1991. *Indian Mushroom* (Nair, M.C. eds.) pp.153-159. Kerala Agricultural University, Vellanikkara.

Purkayastha, R.P. and D.Nayak. 1981. *Mush.Sci.* 9(11): 697-713.

Quimio, T.H. 1993. *Mushroom Research* 2(2):87-90.

Rai, R.D., O.P.Ahlawat and R.N.Verma. 2002. *Recent Advances in the Cultivation Technology of Edible Mushroom* (R.N.Verma and B.Vijay eds) pp.253-259. MSI- NRCM, Chambaghat(HP).

Rainey, P.B., A.L.J.Cole, T.R.Fermor and D.A.Wood. 1996. *Mycological Res.* 94: 191- 195.

Ramsbottom, J. 1953. *Proc.Nutr. Soc.* 12:39-44.

Rangaswami, G. 1978. *Ind. Mush. Sci.* I: 267-274.

Rapeh, J.A. 1976. *J. Gen. Microbiol.* 95: 54-66.

Rasmussen, C. 1962. *Mush. Sci.* 5: 91.

Reddy, P.P., M.S.Rao, R.P.Tiwari and M.Ganesh. 1999. *Pest Management in Hort. Eco. System*. 5: 50-53.

Rinker, D.L., R.Snetsinger and R.Tetrault. 1989. *Mush.Sci.* 12(2): 867-876.

Roy, S., R.C.Anantheswaram and R.B.Beelman 1996. *J.Food Sci.* 61(2): 391-397.

San Antonio, J.P. 1984. *Hort.Reviewers* 6: 85-118.

Seth, P.K. 1980. *Ind.J. Mush.* 6(1): 64-67.

Seth, P.K. and R.L.Munjal. 1981. *Mush.Sci.*11(1): 427-441.

Shandilya, T.R. 1976. *Ind.J. Mush.* 2: 43.48.

Shandilya, T.R. 1988. *Ind.J. Mush.* 14: 20-21.

Shandilya, T.R. 2000. *Ind.J. Mycol. Pl.Pathol.* 30(2): 285.

Sharma, N.K. and A.S.Khanna. 1999. *Diseases of Horticultural Crops, Vegetables, Ornamental and Mushrooms* (Verma, L.R. and R.C.Sharma eds). Indus Publishing Company, New Delhi. Pp.497.

Sharma, S.R. 1997. *Advances in Mushroom Biology and Production* (Rai, R.D.; B.L.Dhar and R.N.Verma eds.). MSI-NRCM, Chambaghat(HP) pp.415.

Sharma, V.P. 1998. *Mushroom Research* 7(1): 1-12.

Sharma, V.P. and C.L.Jandaik. 1995. *Ind.J. Mycol.Pl.Pathol.* 15: 192-194.

Sharma, V.P. and S.R.Sharma. *Ind.J.Mush.* 18: 53-56.

Sinden, J.W. and E.Hauser. 1953. *Mush.Sci.*2: 723.

Singh, Chirag and V.P.Sharma. 2001. *Plant Diseases Research.* 16: 225-229.

Singh, S., C.K.Ganesh and Shrivashraya Singh. 1995. *Indian Food Industry* 14(6): 38- 47.

Sisto, D., S.Faggiano and G.L.Raina. 1997. *Petria* (3): 159-164.

Sohi, H.S., P.K.Seth and S.Kumar. 1965. *Current Sci.* 31: 488-489.

Sohi, H.S., R.C.Upadhyay. 1989. *Mush.Sci.* 12: 611-614.

Song, S.F. 1983. Deptt. of Agriculture, Taiwan Principal Government, Teaching, Taiwan 66pp.

Steiner, G. 1933. J.Agri.Res. 46: 427.

Stoller, B.L. 1969. *The Science and Practice of Mushroom Growing* I: 33-42. Stoller Research Co. Inc.Santacruz, California.

Suman, B.C. 1982. *Ind.J.Mush.* 8: 44-46.

Suman, B.C. 1993. *Ind.J.Mycol.Pl.Pathol* 23(3):314-315.

Suman, B.C. 2003. *Current Vistas in Mushroom Biology and Production* (Upadhyay, R.C., S.K.Singh and R.D.Rai eds.) pp.11-21. MSI-NRCM, Chambaghat, Solan.

Suman, B.C. 2004. *Mushroom Research* 13(2): 54-56.

Suman, B.C. and V.P.Sharma. 2005. *Mushroom Cultivation, Processing and Uses*. Publisher Agrobios(India) Chopasani Road, Jodhpur. Pp.344.

Suzuki, S. and T.Oshima. 1976. *Mush.Sci.* 9(1): 445-462.

Taylor, E. 1992. *Mush.J.* 507: 17-21.

Tewari, R.P. and M.Pandey. 1991. *Indian Phytopathology* 44: 140-143.

Tewari, R.P. and S.J.Singh. 1985. *Indian J.Virol.* I: 35-41.

Thakur, Kiran and S.R.Sharma. 1992. *Mush.Information* 9: 1-10.

Thapa, C.D., C.L.Jandaik and N.K.Sharma. 1987. *J. Mush.Sci.* 2: 169-172.

Thapa, C.D., S.C.Bhardwaj and V.P.Sharma. *Indian Mushroom* (M.C.Nair eds.) pp.61- 62 Kerala Agricultural University. Vellanikkara.

Themas, K.M., T.S.Ramakrishnan and I.L.Narasimhan. 1943. *Madras Agri.* 31: 57-59.

Toulass, A.G. 1915. *Phytopathology* 5: 51-54.

Tourneforte, J. 1707. *Momorois de I, Academic Royale Del Science* 1707: 58-66.

Tresheow, C. 1944. *Dan Bot.Ark.* 61:1

Trivedi, A., S.S.Sharma and A.Doshi. 1991. *Indian Mushroom* (M.C.Nair eds.) pp.166-169. Kerala Agricultural University, Vellanikkara.

Tschierpe, H.J. and J.W.Sinden. 1962. *Mush.Sci.* 5: 61.

Tunney, J. 1978. *Ind.J.Mush.* 4(2): 1-7.

Tunney, J. 1980. *Guidelines and procedure for making bulk pasteurized compost*. Submitted to HPMC, Office Chambaghat, Solan (H.P.) India.

Tu, C.C. and Y.M.Liao. 1989. *Mush.Sci.* 12: 615-622.

Umar, M.H., F.P.Geels and L.J.L.D.Van Griensven. 2000. *Mush.Sci.* 15: 561-567.

Van, der Lende, T.R., M.C.Hermsen and G.H.Vessels. 1994 *Mushroom News*. 20: 4-7.

Van, Zaayen, A. and A.J.Rutijens 1981. *Mush.Sci.* 11: 395-402.

Van, Zaayen, A. and B.Pal-Luiten. 1977. *Netherland, J. Pl.Pathol.* 83: 221-240.

Van, Zaayen, A. W.Gams. 1982. *Netherland J. Pl.Path.* 88: 143-154.

Vedder, P.J.C. 1978. *Modern Mushroom Growing*, Educabook pp.200-420.

Verma, R.N. 2002. *Recent Advances in the Cultivation Technology of Edible Mushroom* (R.N.Verma and B.Vijay eds.) pp.1-10. MSI-NRCM-Solan (HP).

Vijay, B. and S.R.Sharma. 1993. *Ind.J.Mycol.Pl.Pathol.* 23: 121.

Vijay, B. and O.P.Ahlawat. 2002. *Recent Advances in the Cultivation Technology of Edible Mushroom* (R.N.Verma and B.Vijay eds.) pp. 95-132. MSI-NRCM-Solan.

Vogel, F.S., S.J.Mc.Graroy; L.A.K.Kempre and D.G. Graham. 1974. *American J.Pathol.* 76. 165-174.

Wang, Z.S., J.H.Liao and H.C.Wang. 1995. *Mush.Sci.* 13(1): 17-22.

White, P.F. 1995. *Mush.Sci.* 14: 475-84.

Wilkin, W.H.1946. *Br.J.Expt.Pathol.* 28: 416-417.

Wood, D.A. and T.R.Fermor. 1985. *The Biology and Technology of Cultivated Mushroom* (P.B.Flegg, D.M.Spencer and D.A.Wood eds.) pp.43-61. John Willey and Sons, England.

Wright, E.M. and R.J.Chambers 1994. *Entomophaga* 39: 325-335.

Wu, J. and Z.Q.Zhang. 1993. *Expt.Appl.Acasol.* 17(3): 233-240.

Wuest, P.G. and R.K.Moore, 1972. *Phytopathology* 62(2): 1470-1472.

Xu-Jianping, P.A.Horgen and J.B.Anderson. 1996. *Mycological Research* 100(2): 188- 192.

Yadav, M.C. 2002. *Recent Advances in the Cultivation Technology of Edible Mushroom* (R.N.Verma and B.Vijay eds.) pp.35-36. MSI-NRCM-Chambaghat(HP).

Yoshioka, P., T.Sano and T.Ikekawa. 1973. *Chem.Pharm.Bull.* 21: 1772-1776.

Zadrazil, F. 1974. *Mush.Sci.*9(2): 621-652.

Zakia, Bano and M.S.Singh. 1972. *J.FoodSci.Tech.*9(1):18.

Zhi, Y., S.Z.Kang and J.F.Wu 1995. *J. of Shanghai Agricultural College* 13: 129-134.

Index

Plate 1: A View of the Crop of *A. bisporus*

Plate 2: *Agaricus bitorquis*

Plate 3: ***Pleurotus sajor-caju***

Plate 4: ***Volvariella volvacea***

Plate 5: ***Calocybe indica***

Plate 6: ***Lentinula edodes***

Plate 7: *Flammulina velutipes*

Plate 8: *Agrocybe aegerita*

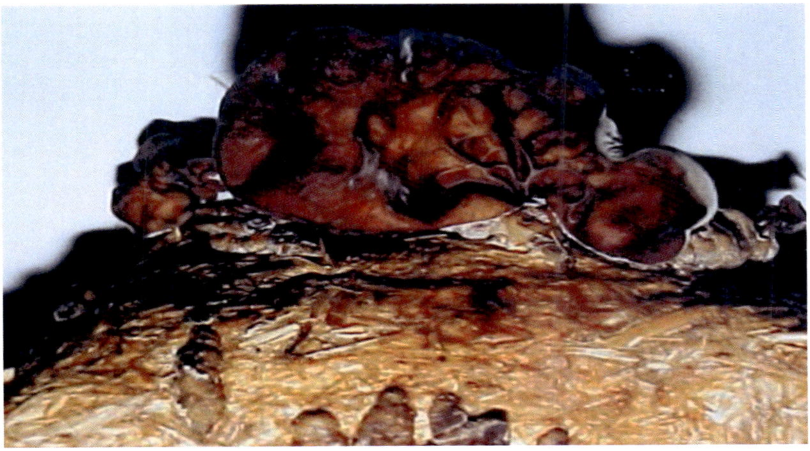

Plate 9: ***Auricularia polytricha***

Plate 10: Symptoms of Wet Bubble

Plate 11: Symptoms of Green Mould

Plate 12: Symptoms of False Truffle

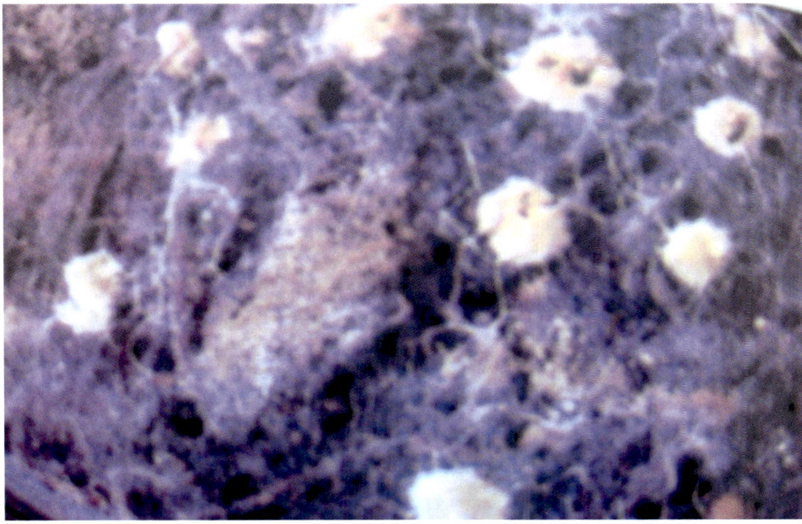

Plate 13: Symptoms of Yellow Mould

Plate 14: Symptoms of Bacterial Blotch

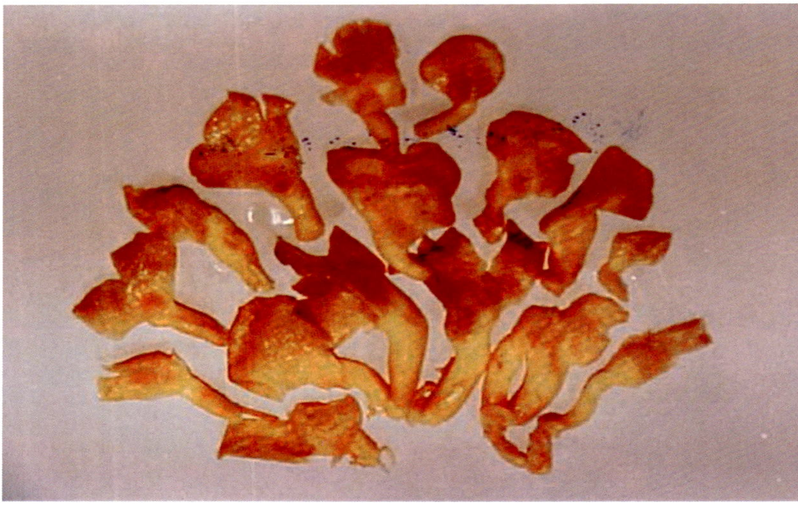

Plate 15: Symptoms of Yellow Blotch

Plate 16: Mushroom cracking